LITHIUM NIOBATE CRYSTALS

Lithium Niobate Crystals

(Physio-chemical Aspects of Technology)

Yu S. Kuz'minov

Viva Books

New Delhi | Mumbai | Chennai | Kolkata | Bengaluru | Hyderabad | Kochi | Guwahati

First Indian Edition 2012

VIVA BOOKS PRIVATE LIMITED

- 4737/23 Ansari Road, Daryaganj, New Delhi 110 002
  E-mail: vivadelhi@vivagroupindia.net, Tel. 42242200
- 76, Service Industries, Shirvane, Sector 1, Nerul, Navi Mumbai 400 706
  E-mail: vivamumbai@vivagroupindia.net, Tel. 27721273, 27721274
- Jamals Fazal Chambers, 26 Greams Road, Chennai 600 006
  E-mail: vivachennai@vivagroupindia.net, Tel. 28290304, 28294241
- B-103, Jindal Towers, 21/1A/3 Darga Road, Kolkata 700 017
  E-mail: vivakolkata@vivagroupindia.net, Tel. 22836381, 22816713
- 7, Sovereign Park Apartments, 56-58, K. R. Road, Basavanagudi, Bangalore 560 004
  E-mail: vivabangalore@vivagroupindia.net, Tel. 26607409, 26607410
- 101-102 Mughal Marc Apartments, 3-4-637 to 641, Narayanguda, Hyderabad 500 029
  E-mail: vivahyderabad@vivagroupindia.net, Tel. 27564481, 27564482
- Beevi Towers, First Floor, SRM Road, Kaloor, Kochi 682 018, Kerala
  E-mail: vivakochi@vivagroupindia.net, Tel: 0484-2403055, 2403056
- 232 GNB Road, Beside UCO Bank, Silpukhuri, Guwahati 781 003
  E-mail: vivaguwahati@vivagroupindia.net, Tel: 0361-2666386

www.vivagroupindia.com

Published by arrangement with
Cambridge International Science Publishing
7, Meadow Walk, Great Abington
Cambridge CB1 6AZ, UK

ISBN : 978-81-309-1237-0

Published by Vinod Vasishtha for Viva Books Private Limited, 4737/23, Ansari Road, New Delhi-110 002. Printed and bound by Syndicate Binders, New Delhi.

FOREWORD

V.S. Osiko, *Academician of the Russian Academy of Sciences*

The lithium niobate crystal has a unique combination of physical properties: it is at the same time a ferroelectric, a piezoelectric and pyroelectric and, in addition to this, has high nonlinear-optical and electro-optical coefficients and photorefractive sensitivity. Because of these properties, these crystals can be used in optical and acoustic devices, and photorefractive sensitivity, enhanced by doping with transition metals, offers new possibilities of using a lithium niobate as a holographic medium for recording information. Because of this wide range of properties, the lithium niobate crystal has been the subject of special attention for almost 30 years.

These properties are determined by the crystallographic structure of lithium niobate which is more sensitive to both physical and chemical effects. This circumstance complicates the production of crystals of high optical quality and uniform chemical composition and the industrial conditions. In this book, small in size, special attention is paid to the physical-chemical special features of technology. Taking these features into account, it is possible to produce crystals satisfying the requirements imposed on them. The disruption of stoichiometry in these crystals and detection of this disruption by physical methods are investigated. At the same time, the concepts and methods proposed in the book are of general nature and can be used in the technology of other crystals.

The book may be useful to experts in the area of single crystal growth and to developers of devices based on lithium niobate crystals, and can also be recommended to students of advanced courses in appropriate specialisation.

About the Author

Prof Yuri Sergeevich Kuz'minov is a Corresponding Member of the Engineering Academy of Sciences of the Russian Federation, Principal Research Associate of the General Physics Institute of the Russian Academy of Sciences, a specialist in the field of nonlinear optical crystals for laser radiation control. He has published more than 150 publications in Russian and foreign journals and three monographs, including Ferroelectric Thin-Film Waveguides in Integrated Optics and Optoelectronics (together with A M Prokhorov and O A Khachaturyan), published in 1996 by Cambridge International Science Publishing.

CONTENTS

Introduction

The aim of this book is to describe more accurately and justify the growth conditions of lithium niobate crystals by technologies used by the majority of crystal producers. However, the author believes that some of these technologies have principal errors which do not make it possible to produce high-quality and chemically homogeneous crystals. The main shortcoming is ineffective or no inspection of the initial reagent as regards the ratio of the main components Li/Nb and secondary impurities. This applies mainly to commercially available lithium niobate powders synthesized by the solid phase method. The producers do not guarantee that the stoichiometry of the reagents is strictly adhered to. The error of determining the chemical composition is $\pm$0.005 mol. At this error, the deviation from the congruent composition, taken from the literature data published 20 years ago ($\alpha = 0.486$ mol) covers a wide range of compositions, including stoichiometric. Isomorphous impurities are most difficult to detect. In the case of lithium niobate, these impurities may include tantalum which accompanies niobium in minerals. The tantalum content in starting niobium pentoxide may reach several percent. Lithium niobate crystals, grown from such a mixture, have higher density which affects certain physical properties, especially the speed of propagation of sound waves in the crystal.

As shown by the authors (Kuz'minov and Osiko, 1992), disruption of stoichiometry in lithium niobate crystals is caused mainly by a shortage of oxygen in the initial niobium pentoxide. Without strict control of the chemical composition of starting components and correction of the starting melt during experimental determination of the congruent composition it is not possible to produce high-quality crystals. Conventional technology also has a number of other significant omissions which also disrupt the chemical composition of the melt. One of them is that a large number of crystals (6–10 boules) are drawn from the same crucible. Even if the congruent composition is determined accurately, there are still small deviations which build up in repeated growth of the crystals without complete replacement of the melt in the crucible.

Relatively large crucibles with a diameter and height up to 150 mm are often used in technology. During operation, the crucibles are greatly deformed and their volume increases. The charge in the crucible is 5–7 kg of lithium niobate powder. 1/5–1/6 of the melt is pulled during crystallization, i.e. a large part of the melt remains in the crucible. The author believes that it is most efficient to use crucibles with a smaller diameter, to 90–100 mm, in which 20–25% of the melt remains after pulling a crystal with a diameter of 76–78 mm and up to 120 mm long. It should be mentioned that reports have been published on pulling more than 90% of melt (Bordui et al, 1991). In this technology cleaning of the crucible is simplified and the degree of deformation of the crucible is considerably lower, i.e. deformation takes place as a result of the increase of the volume of the solidified melt in the crucible. Two or three crystals are pulled from the same crucible which is then completely cleaned. When using a congruent composition melt, this method makes it possible to ensure that the composition remains constant along the length of the crystal boule. For crystals used for producing devices based on surface acoustic waves, this results in a constant rate of propagation of these waves throughout the entire volume of the crystal.

Thus, extensive changes must be made in the technology of crystalline lithium niobate. Without making these changes, the quality of crystals cannot be improved. The physico-chemical properties of the melt and the crystal will be presented and discussed. These properties must be taken into account in developing technology of producing high-quality crystals.

1. Physico-Chemical Properties of the Lithium Niobate Melt

1.1 Temperature dependence of melt viscosity

The physical properties of the melt play a significant role in single crystal growth. The main parameters whose variations indicate the state of the melt are surface tension, viscosity, density and electrical resistivity. These parameters are structure-sensitive and vary even in the presence of small additions. Examination by the methods of combined light scattering shows that melting does not lead to total disruption of the short-range order typical of the solid state (Voron'ko, et al, 1987). Structural transformations in a liquid continue with both increasing temperature and during isothermal holding.

Changes of the short-range order in the structure of the liquid in a specific temperature range are typical of substances whose binding forces greatly change during melting. A short-range order variation indicates that structurally different states of the liquid may exist. The fact that the Curie temperature and melting point of lithium methaniobate are similar indicates that the binding forces greatly change at temperatures higher than the melting point. This in turn causes hysteresis phenomena of the structure-sensitive properties.

To select technological parameters of single crystal synthesis (temperature gradients, pulling rate and crystal rotation), it is necessary to have information on the temperature and time dependence of viscosity, surface tension and melt density. Data on the electrical conductivity of the melt are essential in the process of polarization of the crystal.

The viscosity of a lithium niobate melt was measured in air and helium (Bol'shakov, et al., 1969). The results of measuring kinematic viscosity during isothermal holding in the 'heating regime', Fig.1, show that in all cases the kinematic viscosity varied greatly during the first 5–7 min. After reaching a maximum, the viscosity decreases to a stationary value. A minimum is observed in the cooling regime and the time to its occurrence coincides with that of the maximum. Comparison of the shape of these curves shows that the structure of the lithium niobate melt changes during a period of 10–25 min. It may be expected that

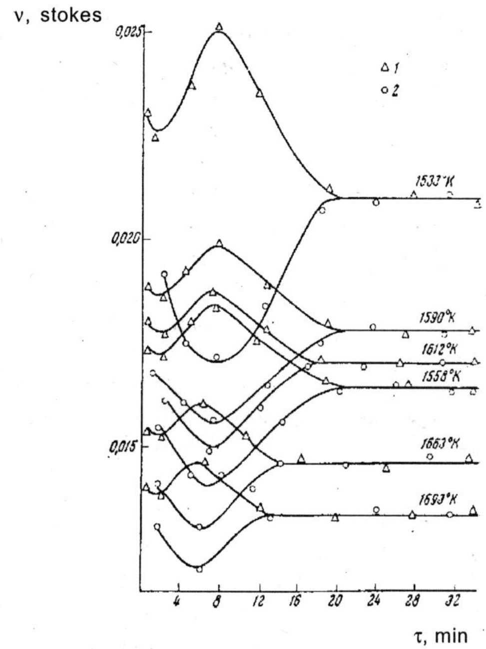

Fig.1 Dependence of the kinematic viscosity of the lithium niobate melt on time in isothermal holding: 1) in heating, 2) in cooling (Bol'shakov et al., 1969).

during rapid heating the liquid will be in the metastable state in which the kinematic viscosity is higher than the equilibrium value.

In rapid cooling, the kinematic viscosity is higher than the stationary value. Thus, it has been established that structural transformations take place in the lithium niobate melt in the temperature range 1560–1590 K over a period of 20–30 min. This period is necessary to stabilize the melt prior to the start of seeding.

The variation of the melt viscosity of the lithium niobate melt in the temperature range 1250–1350°C, in which the lower limit is slightly below the melting point of 1253°C, is shown in Fig.2. Viscosity increases with decreasing temperature. In particular, viscosity increases between 1280 and 1250°C. This temperature range is characterized by temperature fluctuations of the melt in the crucible during crystal growth. The viscosity and density of the melt were measured simultaneously. The temperature dependence of melt density is shown in Fig.3. The density is inversely proportional to the temperature inside the measurement range and the volume expansion coefficient is $\beta = 0.17\times 10^{-3}\ K^{-1}$ (Shigematsu, Anzai, et al, 1987).

1.2 Surface tension

The temperature dependence of the surface tension of a lithium niobate melt

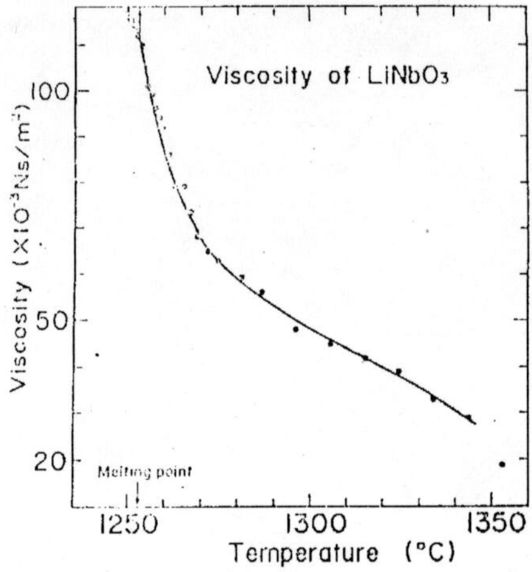

Fig.2 Temperature dependence of the viscosity of the lithium niobate melt (Bordui et al, 1991).

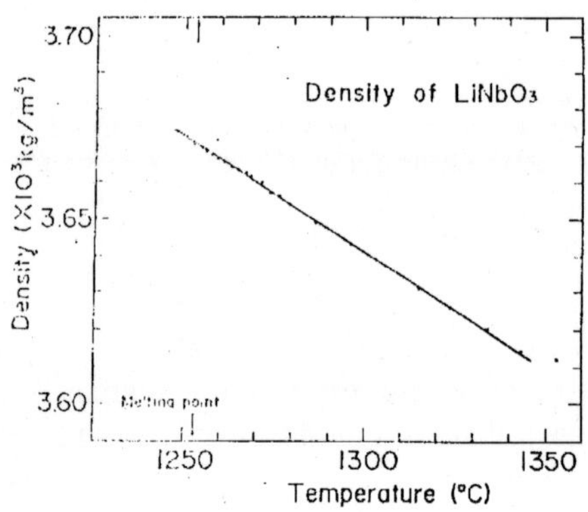

Fig.3 Temperature dependence of the density of the lithium niobate melt (Bordui et al, 1991).

is shown in Fig.4. The error of measurement of surface tension is ± 0.03 N/m. The temperature coefficient of surface tension is $d\gamma / dT = 3 \times 10^{-4}$ N/m K (Shigematzu, Anzai, et al, 1987).

Taking into account the properties of the lithium niobate melt, especially viscosity and surface tension, the authors have assumed that changes may take place in the structure of the melt. In the temperature range between 1280°C and a melting point of 1253°C, the molecules of niobium oxide and lithium oxide are in the form of clusters of a constant shape. At temperatures about 1280°C, there are only traces of the clusters and the melt consists of oxide molecules. This assumption is based on the large increase of viscosity at temperatures below 1280°C and also on an increase of surface tension at temperatures below 1290°C. Further, if it is assumed that the clusters in the melt are in the suspended state in the form of spherical blocks and that the volume of the fraction,

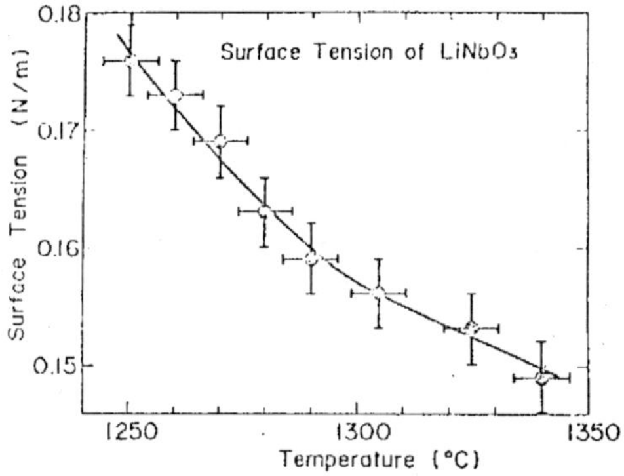

Fig.4 Temperature dependence of surface tension of the lithium niobate melt (Bordui et al, 1991).

occupied by these blocks is less than 30%, it is possible to calculate the volume of the fraction of clusters using the modified Einstein equation for melt viscosity

$$\eta / \eta_0 = 1 + 2.5\varphi + 7\varphi^2 \qquad (1)$$

where η is the actual viscosity, η_0 is the viscosity of the melt consisting of single molecules, and φ is the volume of the fraction of clusters. The calculated volume of the clusters in the melt at 1253°C is 15.5% and at 1260°C it is $\varphi = 8.5\%$. The specified temperature range coincides with the amplitude of thermal fluctuation of the melt during crystal growth. Thus, the volume of the clusters in the lithium niobate melt in the vicinity of the crystal–liquid boundary decreases to 50% of the initial value as a result of high-temperature flows of the liquid which form during thermal fluctuations. It may therefore be assumed that the thermal fluctuations of the melt efficiently reduce the concentration of clusters ahead of the crystallization front and, consequently, cause ordering of the internal structure of the growing crystal. The clusters consist of niobium oxide molecules compacted by lithium oxide molecules. It is possible that tetrahedral clusters were observed by the authors Voron'ko, Kudryavtsev, et al, 1987, who examined a lithium niobate melt by the method of combination light scattering.

1.3 Effect of magnesium oxide on the lithium niobate melt

To increase the phase matching temperature and, consequently, reduce the extent of induced birefringence in crystals, a large amount of magnesium oxide is added to the lithium niobate melt to change the properties of the latter. It is therefore necessary to examine in detail the physico–chemical properties of such a melt.

The density of a lithium niobate melt increases in doping with up to 3 mol% of MgO. However, doping with more than 5 mol% MgO greatly reduces the density. Doping affects not only the density but also the temperature coefficient $d\rho/dT$. High values of $d\rho/dT$ play the role of temperature gradients in the melt and determine the density distribution there. This indicates the existence of two competing mechanisms of variation of the structure of the melt when adding MgO. The first mechanism increases density, the other one reduces it. The first mechanism is obviously more efficient at low MgO concentrations, whereas the second one becomes controlling at a concentration higher than 3 mol%. The details of these mechanisms should be thoroughly examined. The temperature dependence of the density of a lithium niobate melt, doped with MgO, is shown in Fig.5.

The surface tension of the melt (γ) decreases with increasing temperature. The value of γ is the highest for a melt with 3 mol% MgO. The tendency observed in this case is the same as that for the melt density. The temperature dependence of surface tension is shown in Fig.6. A high temperature coefficient $d\gamma/dT$ results in surface convection of the melt by the Marangoni mechanism. When $d\gamma/dT$ increases, surface convection also intensifies. The value of $d\gamma/dT$ increases with increasing level of doping with MgO. When adding 5 mol% of MgO to the melt, the temperature coefficient is more than doubled in comparison with the undoped melt, whereas in doping with 3 mol% MgO $d\gamma/dT$ increases only slightly. Thus, doping with 5 mol% MgO is accompanied by an anomalous behaviour of the lithium niobate melt. These special features of the behaviour of the lithium niobate melt, doped with MgO, must be taken into account in growing lithium niobate crystals with magnesium oxide additions.

1.4 Electrical conductivity

Efficient measurements of the electrical conductivity of the lithium niobate melt can not be carried out because of a low resistance (0.03–0.04 ohm) which represents a small part of the total resistance (0.42–0.43 ohm). The electrical conductivity of the lithium niobate melt is around 10 ohm/m at a temperature of 1250–1340°C. The temperature dependence of electrical conductivity

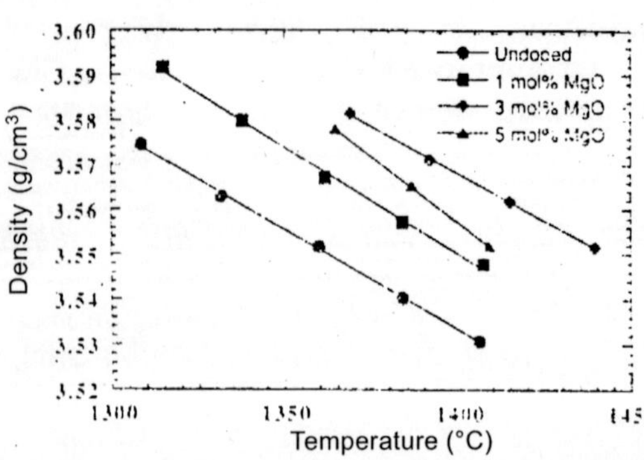

Fig.5 Temperature dependence of the density of the lithium niobate melt doped with MgO (Tokizaki et al, 1992).

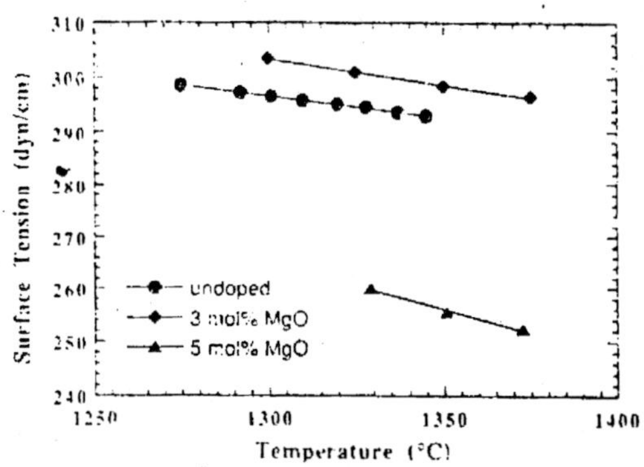

Fig.6 Temperature dependence of the surface tension of the lithium niobate melt doped with MgO (Tokizaki et al, 1992).

was not observed and the frequency dependence is quite weak. The ratio of the electrical conductivity of solid lithium niobate to that of the melt at the melting point is around 40. The frequency dependence of the melt was not observed.

1.5 Congruently melting composition and the distribution coefficient of Li_2O

O'Bryan et al. (1985) used the relationship between the Curie point T_C and the composition of a crystal to determine the Li_2O content of the crystallized melt and also in the upper and lower part of the crystal boule. If it is assumed that mixing in the melt is complete and diffusion in the solid phase may be ignored, the effective coefficient of distribution of Li_2O in the solid–liquid system is determined by the equation $K_0 = K_{ef} = C_S / C_L$, where C_S and C_L is the concentration of Li_2O in the crystal and the melt, respectively. The results for various examined crystals are presented in graphical form in Fig.7. Taking the results into account, it may be seen that the congruent composition ($K_0 = 1.00$) for the lithium niobate corresponds to a concentration between 48.4 and 48.5 mol%. A crystal was grown from a starting melt which contained 48.45 mol% Li_2O. The fraction of the crystallized melt was 0.72. The DTA heating and cooling curves of the starting and the remaining melt, and of the upper and lower part of the boule were used by the authors to determine Curie temperature whose mean value was $T_C = 1138 \pm 2$°C. The congruent composition of the melt was thus determined as 48.45 mol% Li_2O.

Weight losses in the temperature range 1060–1150°C remained constant and equal to approximately 2.5 mg for a crystal weighing 8 g. Weight changes were reversible.

The Li_2O concentration in the crystal is not only the function of the Li_2O content of the melt but also depends on the growth conditions: on the temperature gradients in the melt and growing crystal, on the pulling and rotation rate, convection of the melt and the size of the crucible and the crystal. Grabmaier et al (1991) determined the dependence of the Li_2O content of the crystal on its

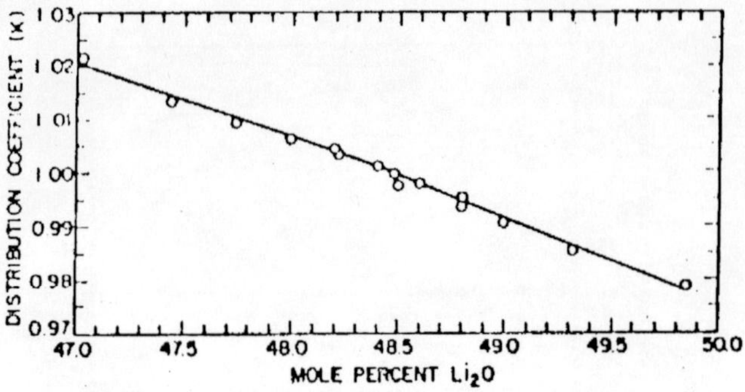

Fig.7 Distribution coefficient of Li in a crystal in relation to the content of Li_2O (mol.%) in the melt (O'Bryan, Gallagher, Brandle, 1985).

content in the melt $\alpha_S = f(\alpha_L)$, shown in Fig.8. The congruent composition of the melt determined in this study was $\alpha_L = 0.483 \pm 0.003$ mol.% Li_2O. This value was in satisfactory agreement with the formula derived by Abrahams and Marsh (Abrahams, Marsh, 1988) $Li_{1-5x}Nb_{1+x}O_3$, $x = 0.0118$ and the experimental data only by Yamada et al. (1987).

In growing lithium niobate crystals, the dependence of the Li_2O content in the crystal on the content in the melt $\alpha_S = f(\alpha_L)$ was initially determined by Carruthers et al. (Carruthers et al., 1971) and much later improved by Grabmaier et al. However, these two dependences, shown in Fig.8, greatly differ. The Grabmaier dependence does not make it possible to produce a crystal of stoichiometric composition, irrespective of the Li_2O content of the melt, whereas according to the Carruthers dependence at $\alpha_L = 0.58$, taking the phase diagram into account (see Fig.9), it is possible to grow a stoichiometric composition crystal. Growth from a double crucible, carried out by Japanese scientists (Kitamura et al., 1982 and Kan, Sakamoto et al., 1992) confirmed this. The Carruthers dependence $\alpha_S = f(\alpha_L)$ was extrapolated using an equation for two concentration ranges of Li_2O in the melt:

$$48.6 < \alpha_L < 58.0$$
$$\alpha_S = 9.817809 + 1.342660\,\alpha_L - 0.011205\,\alpha^2_L \qquad (2)$$

$$58.0 < \alpha_L < 60.5$$
$$\alpha_S = 47.511133 + 0.042911\,\alpha_L \qquad (3)$$

These equations can be used to determine the molar fraction of Li_2O in the crystal from the molar fraction of Li_2O in the melt.

The same authors proposed an equation for evaluating the molar fraction of Li_2O in the melt from the value of the extraordinary refractive index n_e:

$$\alpha_L = 282356.4560 - 256267.118\,n_e + 58157.0995\,n^2_e \qquad (4)$$

The equations presented above may be useful for determining disruptions

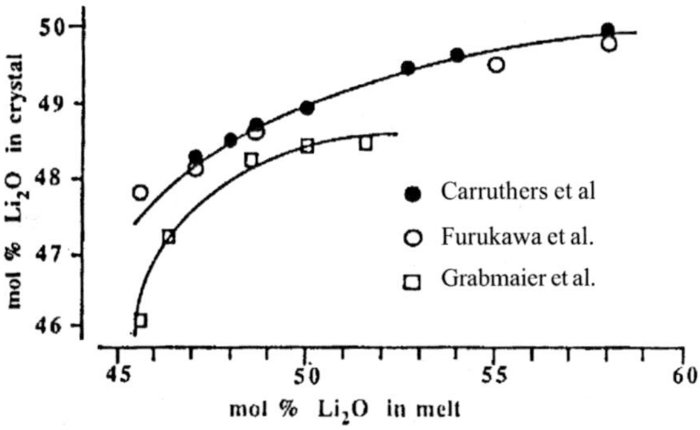

Fig.8 Li content of a crystal as a function of the Li content of the melt (Furukawa, Sato, et al, 1992)

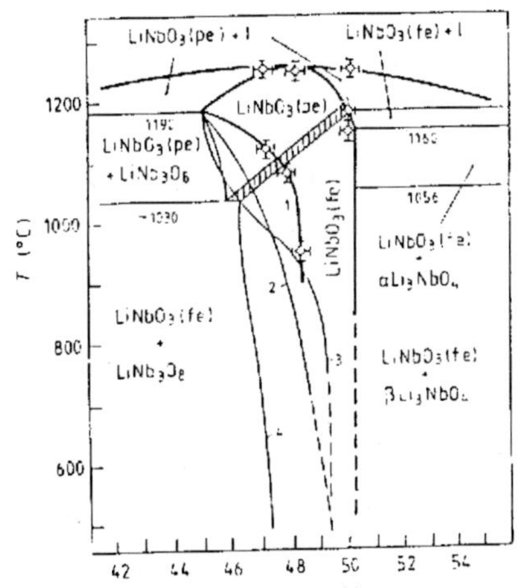

Fig.9 The homogeneity region of $LiNbO_3$, as refined by: 1, Holman (1978); 2, Scott and Burns (1972); 3, Svaasend et al (1974); 4, Lerner et al (1968).

in the stoichiometry of lithium niobate crystals. Problems of the stoichiometry of the melt and crystals will be examined in a considerably greater detail because they are of principal importance for crystal growth. The difference between the dependences derived by Carruthers and Grabmaier is caused, according to the author, by the previously mentioned disruption of the stoichiometry of initial niobium pentoxide.

2. Melt Flows in the Crucible under the Conditions of Induction Heating and Growth of Lithium Niobate Crystals by the Czochralski Method

It is well known that three flows may form in a crucible with a melt in microwave heating and crystal growth by the Czochralski method: smooth movement along a closed curve, Marangoni flows, and forced convection as a result of

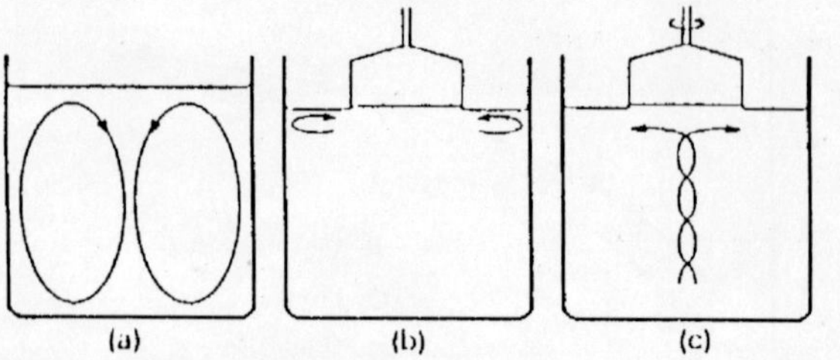

Fig.10 Possible laminar convection of the melt in a crucible in growing the crystal by the Czochralski method: a) free convection; b) Marangoni convection; c) forced convection under the effect of a rotating crystal (Trauth, Grabmaier 1991).

crystal rotation. Figure 10 shows possible forms of laminar flows of the melt in the crucible. Under the effect of typical conditions of crystal growth with microwave heating of the crucible with the melt, forced flows meet smooth convection and Marangoni flows. The flows descend at some distance from the crucible wall. With increasing crystal diameter the forced flow may become controlling with respect to other flows which may generate an inversion of the phase boundary of the crystal.

2.1 Dimensionless quantities describing melt convection

The intensity of convection flows can be described using dimensionless quantities whose expressions are given below.

The characteristic of stationary convective flows is described by the Grashof number $Gr = g\,\beta\,\Delta T^3\,\nu^{-2}$.

The convection on the free surface of the melt is described by the Marangoni number $Ma = -(d\gamma/dT)\ \Delta T\,(R-r)\,(\rho\,\nu\,\alpha)^{-1}$.

Forced convection is described by the Reynolds number $Re = \omega\,r^2\nu^{-1}$. The notations and dimensions of the quantities in these equations are as follows:

Notation	Dimensional representation
g, gravitational acceleration	m/s^2
β, volume expansion coefficient of melt	K^{-1}
$\Delta T = (T_{cruc} - T_{melt})$	K
R, crucible radius	m
ν, kinematic viscosity	m^2/s
$d\gamma/dT$, temperature coefficient of surface tension	N/m K
r, crystal radius	m
ρ, melt density	kg/m^3
α, temperature coefficient of melt diffusion	m^2 K/s
ω, rotational speed of crystal	rad/s

The melt characteristics were determined using the quantities determined from Fig.2,3 and 4 at the melt temperature 1260°C (1533 K), $\beta = 0.17\times10^{-3}$ K^{-1}; $\nu = 90\times10^{-3}$ m^2/s; $d\gamma/dT = 0.1733$ kg m/s^2 K; $\rho = 3.67\times10^3$ kg/m^3, and the parameters of the system were also used $\Delta T = 10$°C, $R = 0.1$ m, $r = 0.066$ m, $\omega = 0.628$ rad/s (6 rpm). Taking these values into account, the Grashof and Reynolds numbers were calculated, $Gr = 2.83\times10^4$ and $Re = 111.55$.

2.2 Inversion of the form of the crystallization surface

Many authors attempted to predict the formation of an inversion of the crystallization front on the basis of results obtained in experiments with growth or physical modelling and also using the ratios of certain growth parameters and material constants. Carruthers (Carruthers, 1976) determined the relationship between smooth rotation and forced convection using the relation $Gr = Re^2$. This relation is in satisfactory agreement with the experimental data obtained by Takagi and Fukazawa (1976). Kobayashi (1981) carried out numerical modelling of the hydrodynamic processes in oxide melts. He found that an inversion forms if the ratio $Re^{2.5}/Gr$ is in the range up to 10.

Other authors (Trauth and Grabmaier, 1991) concluded that the form of the interphase boundary does not depend on the height of the melt in the crucible in a specific range of the height of the melt/crucible diameter ratio. The crystal cannot grow below the critical value of this ratio. The form of the interphase boundary is determined strictly by the rotation rate of the crystal. Initially, all crystals have the concave form of the surface. The surface curvature decreases with increasing crystal diameter. As the rotation rate of the crystal increases, the surface form is straightened towards the edge of the crystal. At a rotation rate of 7.5 rpm in the region below the zone of expansion of a crystal with a diameter of 4" the convex form of the surface changes to wave-like. Above 10 rpm a reversed inversion of the surface form takes place – from convex to concave. At 10 rpm, the surface is sinusoidal (convex in the vicinity of the edge and concave in the centre). At 20 and 30 rpm, a concave surface with a large curvature appears. Inversion is transformed to opposite (changes the sign) if the rotation rate is reduced below the critical value. At a smaller crystal diameter, inversion takes place at a higher rotation rate of the crystal, i.e. N_c is inversely proportional to diameter d. Figure 11 shows the typical pattern of inversion of the crystallization front of lithium niobate. For crystals with a diameter in the range 2–8 cm the critical rotation rate of the crystal under the experiment conditions (Trauth, Grabmaier, 1991) is described by the equation

$$N_c = -4.1d - 41 \pm 2 \qquad (5)$$

where N_c is in rpm, d in cm. This dependence is shown in graphical form in Fig.12.

Brice (Brice, 1968) predicted that melt flows in the crucible may become unstable if the rotation rate exceeds the maximum value $\omega_C > Re_c \eta/a^2$, where

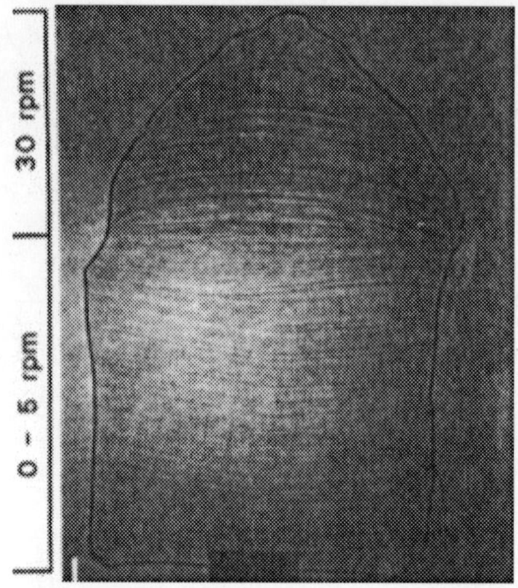

Fig.11 Form of the crystal–melt surface (rotation speed of the crystal 0.30 rpm) (Trauth, Grabmaier 1991).

Re_c is the critical Reynolds number, η is melt viscosity and *a* is the crystal diameter.

A stability criterion of a flat crystallization front was proposed by Chernov (Chernov, 1980): $G_T/V > -mG_c/D$, where G_T is the temperature gradient at the crystallization front, V is the crystal growth rate, G_c is the concentration gradient of the impurity in the melt in the vicinity of the interphase boundary, $C_c = C_0(1-K)/K$, C_0 is the impurity concentration in the volume of the melt, $K=C_S/C_L$ is the impurity distribution coefficient.

To determine the conditions under which there is no inversion of the form of the phase surface, we shall use the Carruthers relationship $Gr = Re^2$. For the crystal growth conditions considered here, we have an inequality $2.83\times10^4 > 1.24\times10^4$. For the Kabayashi condition $Re^{2.3}/Ge \cong 10$, substituting the numerical values we obtain $13.14\times10^4/2.83\times10^4 = 4.64$, i.e. it is not fulfilled. The condition of the flat solidification front according to Brice $\omega < Re\ \nu/a^2$, the numerical values lead to the inequality $\omega < 11\ s^{-1}$. It should be mentioned that the crystal growth conditions selected here are not completely optimum but there are two degrees of freedom: the variation of temperature difference between the crucible and the melt and the ratio of the radii of the crystal and the crucible.

The critical rotation rates in relation to the crystal diameter, determined by experiments (Trauth, Grabmaier, 1991), are shown in Fig. 12.

2.3 Temperature fluctuations in the melt

Iliev et al (1991) divided the total range of the rotation rate of the crystal into three partial ranges in relation to the nature of temperature fluctuations of the

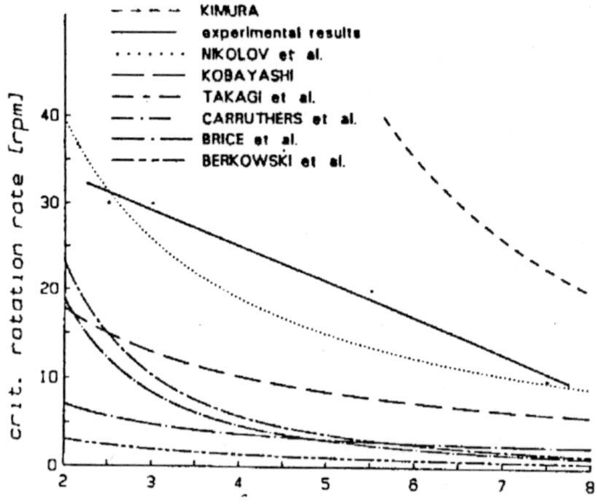

Fig.12 Critical rotation rate in relation to the diameter of the lithium niobate crystal (Trauth, Grabmaier, 1991).

melt: 1. from 0 to N_1, in which aperiodic temperature fluctuations are unstable; 2. from N_1 to N_2 in which there are periodic fluctuations with periods and amplitudes typical of every partial case (in this range the amplitudes show a weak dependence on the rotation rate); 3. above N_2, in which the periods and amplitudes rapidly decrease. The values of N_1 and N_2 depend on the initial state of the system (geometry and Grashof number) and also on the position of the thermocouple in the melt. A high value of Gr and a large distance between the thermocouple and the central part of the melt result in high values of N_1 and N_2. The above-mentioned variations of the nature of thermal fluctuations are determined by the variation of the form of convective flows and by increasing rotation rate. The results show that aperiodic fluctuations are recorded by the thermocouple in a freely convective flow when the rotation rate is in the range $0 - N_1$. The periodic fluctuations are recorded if the thermocouple is situated in the boundary layer between free and forced convective flows and is influenced by the periodic primary thermal and secondary cold flows (rotation rate between N_1 and N_2).

If the rotation rate reaches values at which the thermocouple is always in the forced convection flow with a constant speed, the fluctuations are not recorded. The results show that the transition from aperiodic to periodic fluctuations is accompanied by a relatively large reduction of the measured mean temperature. This shows that the periodic temperature fluctuations in the melt of real oxide systems are closed linked, mainly with the interaction between two convection flows. It should be mentioned that these periodic fluctuations are accompanied by fluctuations of a different nature associated with deviations between the axes of the crucible and the crystal and also with the nature of temperature control.

The fluctuations which are important for crystal growth are observed in the immediate vicinity of the crystal–melt surface. The results obtained in simulation investigations by Iliev et al (1991) show that the range of the rotation rate of the crystal between $N_1 = (0.4–0.5)N_c$ and $N_2 = (1.0 - 1.1)\,N_c$ is an essential condition for the formation of periodic temperature fluctuations below the crystal surface. Here N_c is the critical rotation rate of the crystal at which the forced convective flow reaches the edge of the crystal. As shown by Nikolov et al (1988), at $N = N_c$ the crystal–melt surface becomes flat. The value of N_c is determined using the equation

$$N_c = 3.18\,(g\,\beta\,\Delta T)^{0.44}\,D^{0.245}\,h^{0.155}\,\nu^{0.12}\,d^{-1.08} \tag{6}$$

Simulation experiments show that the value of the Grashof number Gr has a strong effect on the maximum period of temperature fluctuations. Figure 13 shows that the fluctuation period increases with increasing crucible diameter. At the same time, the maximum amplitude of fluctuations depends mainly on the temperature difference between the crucible wall and the surface of the growing crystal and depends only slightly on the Grashof number (Fig.14).

3. Anisotropy of the Growth of Lithium Niobate Crystals

The anisotropy of growth of lithium niobate crystals was studied by the method of in situ examination at a high temperature (Jin, Tsukamoto, 1992). This method makes it possible to examine the transformation of a smooth surface to a cellular one during crystal growth. The experiments were carried out on thin crystal sheets 50–100 μm thick which were cut from oriented crystals of stoichiometric lithium niobate. Sheets with orientation $(2\bar{1}\bar{1}0)$ and $(10\bar{1}4)$ were examined. Pieces of the sheets were placed on a heating loop and gradually heated to melting. The loop diameter was around 6 mm and the temperature difference between the central part and the edge of the specimen was 5–10°C, with the exception of areas close to the heater. The stability of the melt temperature was maintained at $\pm$0.3°C on a level of 1300°C. Under these conditions, the lithium niobate melt was held in the loop by means of surface tension forces. A small piece of the crystal with a specific orientation was immersed into the melt and held in it for a specific period of time at constant temperature. This seed crystal was used for growth experiments in precision control of undercooling. The growth process was examined in a microscope in transmitted light and recorded with a videocamera.

At small undercooling, the smooth surface grew at the periphery of the crystal sheets with orientation $(2\bar{1}\bar{1}0)$and $(10\bar{1}4)$. The results of measurements of the displacement of the circular surface parallel to each other in the direction of the normal between the consecutive growth stages were used to determine the growth rate of the crystal in relation to undercooling temperature. The dependence shown in Fig.15 is linear. With a further increase of the degree of undercooling depressions formed on the smooth growth surface. In particular, in the sheet

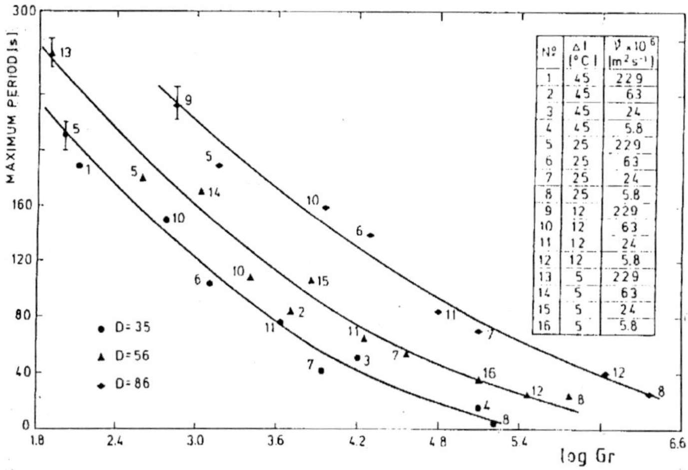

Fig.13 Dependence of the maximum period of temperature fluctuations on Grashof number for crucible of different diameters (Iliev et al, 1991).

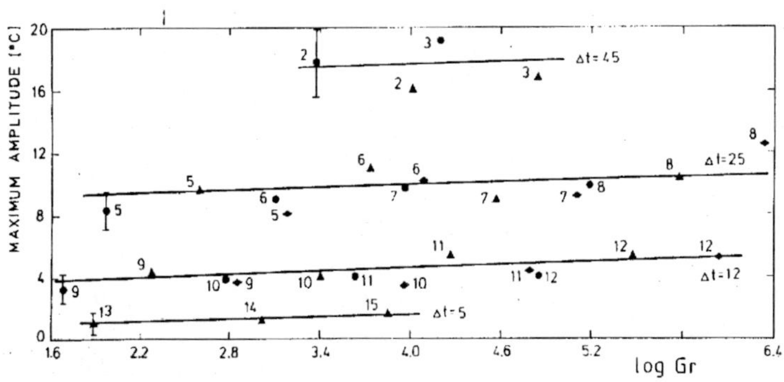

Fig.14 Maximum amplitude of temperature fluctuation for different values of the temperature difference between the crucible walls and the crystal (Iliev et al, 1991).

with orientation $(2\bar{1}\bar{1}0)$ depressions formed at undercooling $\Delta T = 61°C$.

Although the smooth surface and depressions coexist, the faceted growth forms may not be observed. The stage in which smooth sections and depressions coexist may be regarded as the critical stage of undercooling. With increasing undercooling the smooth growth surface transfers to a cellular structure. Figure 16 shows the consecutive stages of growth of a cellular structure on the $(2\bar{1}\bar{1}0)$ face at undercooling $\Delta T = 89°C$. As shown in Fig.16a,b, in the initial stage the smooth surface is unstable in the sense of formation of a saw-like coarse surface. This form of the surface is a consequence of the formation of almost rectangular cells, Fig.16c,d. In earlier experiments (Jin W., Jin J, et al, 1987) it was shown that the faces of rectangular cells are formed by close-packed layers $\{01\bar{1}2\}$.

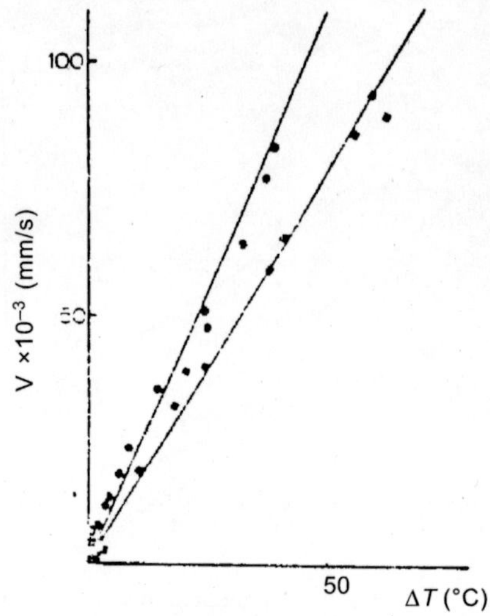

Fig.15 Variation of growth rate (V) of the smooth interface parallel to itself along the normal direction with supercooling for the periphery of the $(2\bar{1}\bar{1}0)$ and $(10\bar{1}4)$ seed crystal: (■) $(2\bar{1}\bar{1}0)$ seed crystal, (●) $(10\bar{1}4)$ seed crystal (Jin, Tsukamoto, 1992).

Analysis of the photographs of consecutive stages of the development of the growth process at high undercooling of the melt shows the presence of the terraced (horizontal) form of growth along the $\{01\bar{1}2\}$ face, being the dominant factor in this growth process. Growth in the direction normal to the $\{01\bar{1}2\}$ face takes place by means of the combination of a large number of horizontal growth layers. Regardless of the differences in the orientation of the seed crystal, a large number of $\{01\bar{1}2\}$ faces always form on the growth surface. It is evident that this phenomenon is associated with the internal properties of the transition process from the smooth surface to the surface with a cellular structure. A similar morphology was observed in impurity bands of rotation of lithium niobate crystals grown by the Czochralski method.

4. Effect of Growth Parameters on Crystallization of Lithium Niobate

If a single crystal grows from the melt, the quality of this crystal depends on a number of technological parameters which determine morphological stability, homogeneity and absence of defects. Morphological stability depends on composition, form and stability of temperature between the melt and the growing crystal. These aspects of quality are determined to a large extent by the crystallization mechanisms at the phase boundary and the results of many investigations show that it is necessary to control this stage.

Lee et al. (Lee, Kim et al., 1992) examined the effect of vertical temperature gradients (G), the crystal growth rate (V) and convection in the melt on the form and stability of the crystal/melt surface during the growth of a lithium

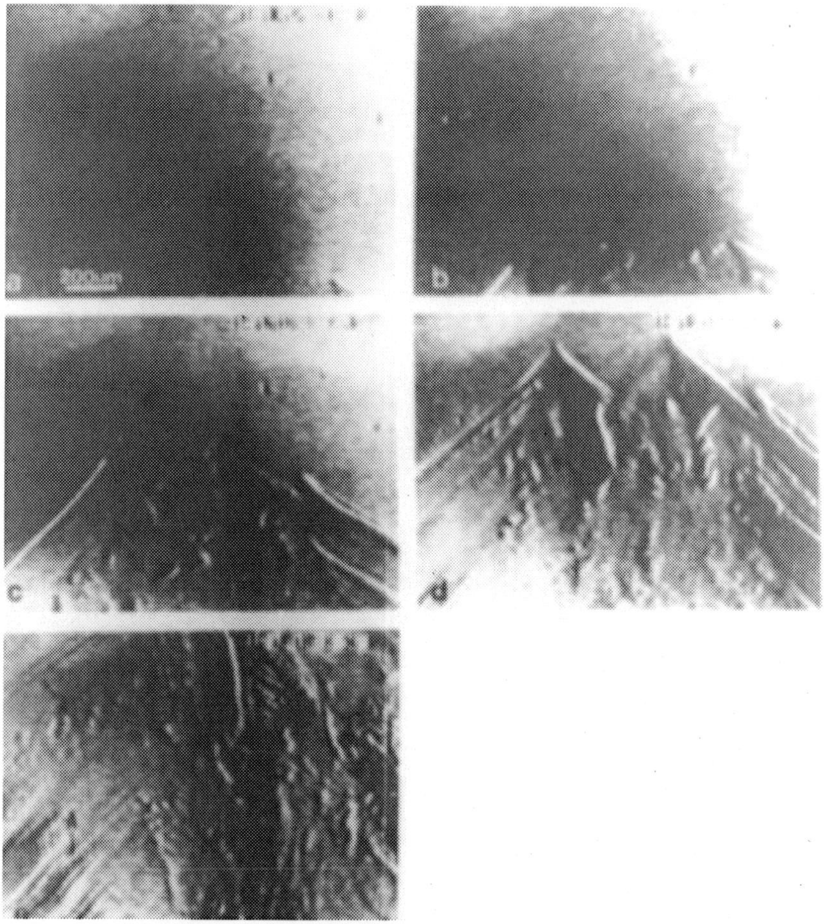

Fig.16 Consecutive stage of growth at the periphery of the $(2\bar{1}\bar{1}0)$ plane of the seed crystal at $T = 89°C$: a) $t = 0$ s; b) $t = 0.7$ s; c) $t = 1.9$ s; d) $t = 2.9$s; e) the shape of impurity striation (Jin, Tsukamoto, 1992).

niobate crystal. The effect of temperature gradients on the quality of the crystal is summarized in Table 1. With a vertical temperature gradient of $G = 22$ K/mm, cracks form during cooling of the crystal and their formation rate is directly proportional to the temperature gradient. At $G = 5$ K/mm, no cracks form in the crystal. However, at a temperature gradient smaller than 3 K/mm the form of the crystal is distorted.

Examination of the effect of the crystal growth rate (V) shows that when the rate is increased above 8.3×10^{-3} mm/s the crystal contains a large number of cracks induced by residual thermal stresses. A very large number of cracks forms in the lower part of the crystal. At rates lower than 2.8×10^{-3} mm/s crystals free from cracks can be produced. Figure 17 shows the crack formation conditions at different values of the parameters G and V. Depending on the values of G and V, there are ranges of rapid crack formation, the range in which cracks form, and the region in which no crack appears. However, at a favourable combination of the values of G and V there are problems with controlling the crystal diameter.

To control the crystal diameter, in addition to G and V it is also important to take into account other crystal growth parameters, especially the rotation rate (N). At a constant temperature gradient and at a growth rate of 4 K/mm and 1.4×10^{-3} mm/s, respectively, Lee et al. determined by experiments the form of the phase boundary in relation to the rotation rate of a crystal 80 mm in diameter (Fig.18). At $N = 6$ rpm the surface was conical, at 8 rpm it was flat, and when the rotation rate was increased to 12 rpm the surface was bent.

When growing a crystal by the Czochralski method with a flat crystallization surface the conditions are favourable for the crystal to grow without cracks. However, cracks can form during cooling of the crystal. At a vertical temperature gradient of less than 4 K/mm twinning of the crystal takes place. Therefore, the optimum temperature gradient for growing lithium niobate crystals 80 mm in diameter with no cracks and twins should be higher than 4 K/mm, in accordance with Fig. 17.

When growing crystals by the Czochralski method, convection of the melt in the crucible is determined by the rotation rate of the crystal (N). In the case in which $N < N_c$, where N_c is the critical rotation rate at which the crystal has a flat crystallization front, the temperature difference ΔT between the crucible wall and the melt/crystal surface causes forced bottom convection along the axis of the crucible below the crystal. If $N > N_c$, upper forced convection takes place.

As already mentioned, the form of the surface and its stability during crystal growth have a controlling effect on crystal quality. For example, if a crystal is conical, cracks, facets and dislocations can form quite easily because in this case it is difficult to ensure that the diameter of the growing crystal remains constant. It is therefore necessary to find a relationship between the parameters G, V and N in order to calculate the critical rotation rate of the crystal (N_c). This speed corresponds to the flat crystal/melt surface. These problems will be examined later. Using the dimensionless Grashof and Reynolds numbers for the characteristics of the melt in the crucible, Nikolov et al. (Nikolov

Table 1 Effect of the vertical temperature gradient on growth of a lithium niobate single crystal with a diameter of 25 mm (Lee, Kim, et al, 1992)

Vertical temperature gradient (K/mm)	Growth rate V, (10^{-3} mm/s)	Cracks
5.0	2.78	no
15.6	2.92	small
21.6	3.22	medium

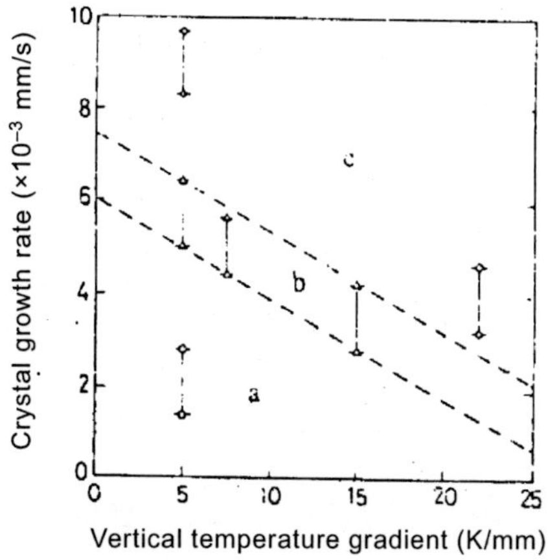

Fig.17 Diagram of vertical temperature gradient *vs* growth rate showing crack formation in $LiNbO_3$ single crystal of 25 mm diameter: a) crack-free from cracks; b) crack-forming region; c) severe crack region (Lee et al, 1992).

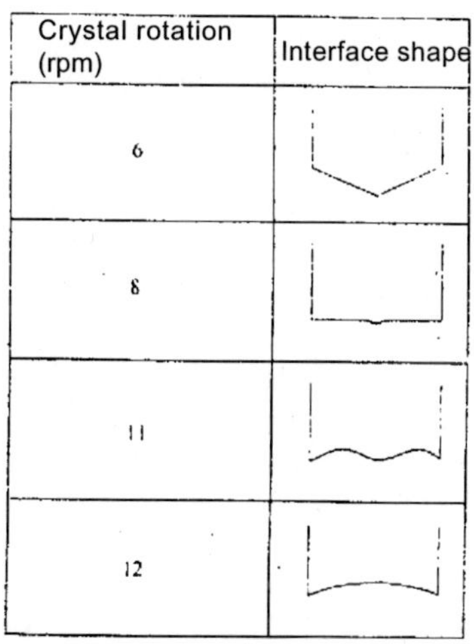

Fig.18 Variations of the interface shape in relation to the rotation rate during growth of a LN single crystal 80 mm in diameter and 132 mm long (Lee et al, 1992).

et al., 1988) derived empirically an expression for the rate of bottom forced convection:

$$W_d d / v = 0.0129\left[\mathrm{Gr}(d/D)^{2.1}(h/D)^{0.35}\right]^{0.73} \tag{7}$$

The rate of upper forced convection (W_1) in the direction of the crucible axis is expressed as a function of the Reynolds number (Re) and the crucible diameter (D) by the following equation:

$$W_1\, d/\nu = 0.002\ \mathrm{Re}^{1.6} \text{ for Re} < 75 \pm 4 \tag{8}$$

$$W_1 d/\nu = (0.072\text{–}0.006\, D)\ \mathrm{Re}^{0.73+0.04D} \text{ for Re} > 75 \pm 4 \tag{9}$$

where D is measured in cm.

The Reynolds number for the lithium niobate crystal 80 mm in diameter was Re < 75 and, consequently, equation (8) can be used. The conditions corresponding to the flat surface can be determined from equation $W_d = W_1$. Combining equations (7) and (8), we derive an equation for determining the critical rotation rate of the crystal at which a flat phase boundary surface forms:

$$N_c = 3.20\ (g\beta\Delta T)^{0.45}\ D^{0.25}\ h^{0.16}\ \nu^{0.10}\ d^{-1.06} \tag{10}$$

This expression slightly differs from that given previously and it is evident that it satisfies more efficiently the experimental conditions of the authors (Lee, Kim et al., 1992).

Table 2 summarizes the experimental conditions for growing single crystals of lithium niobate 25 and 80 mm in diameter by the Czochralski method. If the crystal is 132 mm long, the height of the melt in the crucible is 42 mm and $\Delta T = 10$ K, then using the values $\beta = 1.5\ \mathrm{K}^{-1}\ \mathrm{mm}^2/\mathrm{s}$ and $v = 1.5$ mm/s (Trauth et al, 1991), it is possible to calculate the critical speed which is 7.6 rpm. This value is in agreement with the experimental results. Thus, by controlling the rotation rate of the crystal, it is possible to grow crystals of large diameters.

5. Effect of Impurities on Crystallization of Lithium Niobate

Bridenbaugh (Bridenbaugh, 1973) was one of the first scientists who paid attention to the effect of impurities introduced into the melt (of the order of molar percent) on the variation of the ratio of the main components in the crystal growing in the melt. An impetus for these investigations was the need to increase the temperature of phase synchronism for second harmonic generation (T_{pm}). Examination of the Li_2O–Nb_2O_5–oxide ternary phase diagram shows that an addition of 1% oxide to a congruent melt of lithium niobate results in an excess of approximately 0.6 mol% of Li_2O in the melt, if the slope of the liquidus line follows the constant Nb_2O_5 isopleth. The equilibrium distribution constant for the excess of Li_2O in the melt, determined by Carruthers, is $K_0 =$ 0.3. The most suitable oxide is MgO which greatly increases T_{pm} and the melt remains congruent. This is very important for technology of nonlinear optics crystals.

If the melt contains an impurity, it is always necessary to examine its distribution in the crystal volume. Therefore, Bridenbaugh attempted to measure the effective distribution coefficient K_{ef} for different growth rates and temperature gradients in the melt. These data together with the values obtained in examining the growth rate at which the phase boundary is disrupted and a con-

Table 2 Growth conditions for lithium niobate single crystals (Lee, Kim et al, 1992)

	Crucible	
Material and purity	Pt 99.995%	Pt 99.995%
Diameter, D (mm)	44	126
Height, H (mm)	44	130
Thickness, t (mm)	0.4	2
	Crystal	
Diameter, d (mm)	25	80
Length, l (mm)	45	150
Vertical temperature gradient, G (K/mm)	3, 5, 16, 22	4
Crystal growth rate·, V (mm/s)	$(1.4–10)\times10^{-3}$	1.4×10^{-3}
Rottational speed of crystal, N (rpm)	40	6–12
Cooling rate (K/h)	80	50

centration undercooling appears can be substituted into the Burton–Prim–Slichter equation (Burton, et al., 1953, Hurle 1961) to determine the diffusion constant D and the thickness of the diffusion layer δ.

Concentrational undercooling was observed at a growth rate of 1.25×10^{-3} m/s. This value, together with the measured effective distribution coefficient of the impurity represented by the excess content of Li_2O above the congruent composition $K_{ef} = 0.45$, were included in the equation derived by Burton et al (1953) to determine the value $\Delta = R\delta / D$, included in the Burton equation:

$$K_{ef} = K_0 / [K_0 + (1 - K_0)e^{-\Delta} \quad (11)$$

Here K_0 is the equilibrium distribution coefficient, R is the growth rate, δ is the thickness of the diffusion boundary layer, and D is the diffusion coefficient. As a result of substitution we obtain $e^{-\Delta} = 0.408$ and $\Delta = 0.90$.

To determine the diffusion coefficient, Bridenbaugh used the equation derived by Hurle (Hurle, 1961), which includes the physico–chemical constants of the melt:

$$G_L = [-m R c_L / D](\Gamma + e^{-\Delta}) \quad (12)$$

where G_L is the temperature gradient at the phase boundary, c_L is the impurity concentration, m is the angle of inclination of the solidus on the phase diagram, R is the growth rate of the crystal, and $\Gamma = K_0/(1 - K_0)$, where K_0 is the equilibrium coefficient of impurity distribution. As regards the numerical values of the quantities included in the equation, it should be mentioned that they were determined for a lithium niobate crystal with a cross section of 1 cm^2, grown from a stoichiometric melt. The volume of the crucible was 30 cm^3, G_L = 150°C/cm; $e^{-\Delta}$ = 0.408; c_L = 1.4 mol% Li_2O (the difference of Li_2O concentrations between the stoichiometric and congruent composition of the melt); m = 3.18°C/mol% Li_2O; K_0 = 0.25 (Carruthers et al., 1971). Consequently, the diffusion coefficient determined in this manner is $D = 5.02\times10^{-5}$ cm^2/s. Remembering that $\Delta = 0.90$, the thickness of the diffusion boundary layer can be calculated from $\Delta = R\delta/D$ and, consequently, δ = 0.036 cm. In turn, the thickness of the diffusion boundary layer is linked with the rotation rate by the equation (Bridenbaugh, 1973): $\delta = C\,\omega^{-1/2}$, where ω is the rotation rate of the crystal and C is a constant which depends on the diffusion coefficient and the kinetic viscosity of the melt. Using the previously determined value of δ and the rotation rate of 30 rpm, we obtain $C = 0.198$. Thus, $\delta = 0.198\omega^{-1/2}$. Using the results, Bridenbaugh plotted the dependence of K_{ef} on $\Delta = R\delta/D$, shown in Fig.19, which can be used to determine the effective distribution coefficient of the impurity in the crystal.

6. Inclusions of Hydroxyl Groups in Lithium Niobate Crystals

One of the secondary ions, found constantly in lithium niobate crystals, are ions of hydroxyl groups OH^- which come from the melt produced by melting a moist initial powder and also when the polarization process is carried out in a humid atmosphere (Zharikov, Ivleva, Kuz'minov, 1975).

OH^- ions are detected on absorption spectra in the infrared region of the spectrum in the vicinity of 2.87 μm consisting of two absorption bands at 2.865 and 2.875 μm. The two absorption bands correspond to different positions occupied by the hydroxyl groups in the lithium niobate lattice. The crystals grown un-

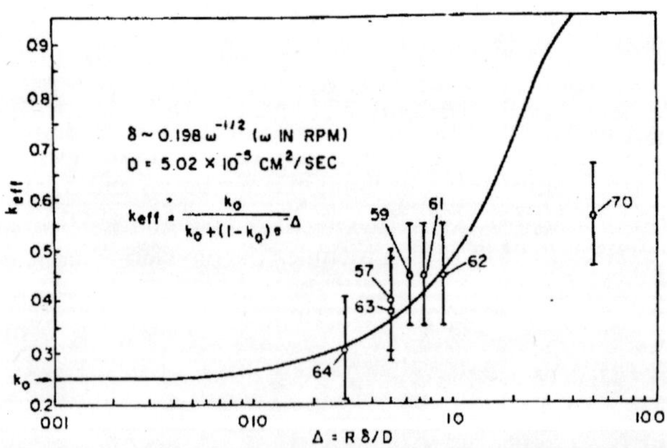

Fig.19 Dependence of the effective coefficient of distribution of Li_2O in a lithium niobate crystal, grown from a stoichiometric melt, on the value of $R\delta/D$ (Bridenbaugh, 1973).

der normal conditions have absorption coefficients at a wavelength of 2.87 μm $K = 2–3\ cm^{-1}$. As already mentioned, the content of OH^- ions increases when annealing the crystals in the atmosphere of water vapour. At a pressure of 10–30 atm and a temperature of approximately 750°C, the absorption coefficient is higher than 30 cm^{-1}. The crystals were annealed in sealed quartz vessels.

The OH^- concentration can be evaluated using the relationships:

$$n\,(OH^-) = (3 \pm 1)\times 10^{19}\, k\,(2.87\ \mu m)\ cm^{-3},$$
$$c\,(OH^-) = (5.0 \pm 1.7)\times 10^{-5}\, k\,(2.87\ \mu m)\ mol,$$

where k is the absorption coefficient at the maximum at a wavelength of $\lambda =$ 2.87 μm, and n and c are the concentration and molar fraction of the OH^- ions, respectively, obtained in examining the resonance of protons (Bollmann, Schilothauer, Zogal, 1977).

The results of investigations carried out by Bollmann et al. (Bollmann, Stohr, 1977) on lithium niobate crystals with different ratios of components R_k = Li/Nb show that the position of the maxima of the absorption coefficients correspond to

$$\lambda_{max} = 2.865\ \mu m \text{ for } R_k < 1$$
$$\lambda_{max} = 2.875\ \mu m \text{ for } R_k > 1$$

The half-width of the absorption band $\lambda = 2.865$ μm is $H = (30.3 \pm 1.4)$ $cm^{-1} = (3.76 \pm 0.17) \times 10^{-3}$ eV. The crystals close to the stoichiometric composition $R_k \sim 1$ usually show mainly the absorption band at 2.865 μm and weak absorption at a wavelength of 2.875 μm. There are literature data showing that this behaviour is also observed in crystals with congruent composition (Herrington et al, 1973). However, Ballmann detected only traces of the band at 2.875 μm in congruent crystals. Annealing in water vapours resulted not only in broadening of the band at 2.87 μm but also in the appearance of a band at 2.25 μm which splits into bands at 2.248 and 2.255 μm. The splitting of the 2.25 μm band corresponds to the splitting of the 2.87 μm band, so that it can be concluded that the absorption band in the vicinity of 2.25 μm is also caused by the presence of hydroxyl groups in the crystal. The absorption spectra of the crystals for various ratios R_k are presented in Fig.20.

Examination of the electrical conductivity σ of the lithium niobate crystals shows that σ increases with increasing OH^- content the crystal, Fig.21. There is a linear relationship between σ and the absorption factor K (2.865) at $R_k <$ 1, if it is possible to prevent reduction of the crystal during its growth leading to the appearance of electronic conductivity. The linear dependence shown in Fig.21 indicates that the conductivity is caused by the OH^- ions and the enthalpy of activation of conductivity is the activation of mobility $\mu(OH^-)$ of OH^- ions. For the absorption band at 2.865 μm, which corresponds to the substitution of positions of O^{--} ions by the OH_0^- ions, the activation energy is: $h_B = (1.12 \pm 0.03)$ eV. The hydroxyl groups, occupying interstitial positions OH^-_i with the 2.875

μm absorption band, have the activation energy equal to $h_B = (1.08 \pm 0.02)$ eV. In the study by Ballmann et al. (Ballmann, Stohr, 1977), the authors published equations for calculating the mobility μ, the diffusion coefficient and jump frequency $v = \nu kT / e\bar{r}^2$ ($\bar{r} = 2.3$ Å is the average jump distance) for hydroxyl groups in oxygen and interstitial positions.

The jump conductivity is dominant in the conductivity dependence of lithium niobate crystals. It is linked with the vibrational frequency of the lattice by the equation:

$$\nu = \nu_0 \exp(s_B / k) \exp(-h_B / kT) \qquad (13)$$

The vibrational frequency of the crystal lattice ν_0 of the examined ions may be approximated by the Debye frequency ν_D, i.e. the latter is determined mainly by the type of ions which are capable of changing their positions with increasing temperature. In the case of lithium niobate crystals, only O^{--} ions are capable of changing their positions at $T < T_0$. As a result of the fact that the masses of the OH^- and O^{--} ions are similar, the vibrational frequency of the lattice may be represented by the Debye frequency for evaluating $s_B(OH^-)$. The available values of the Debye temperature $T_D = (571–583)$ K (Peterson, Carruthers, Carnevale, 1970) were used to determine the values $\nu_D = 1.2 \times 10^{12}$ s^{-1}; $s_B = (OH_0^-) = 1.7\ k$ and $s_B = (OH_1^-) = 2.5\ k$.

In crystal technology, it is important to have information on the mechanisms of implantation of OH ions into the crystal lattice, i.e. the OH^- ions, implanted

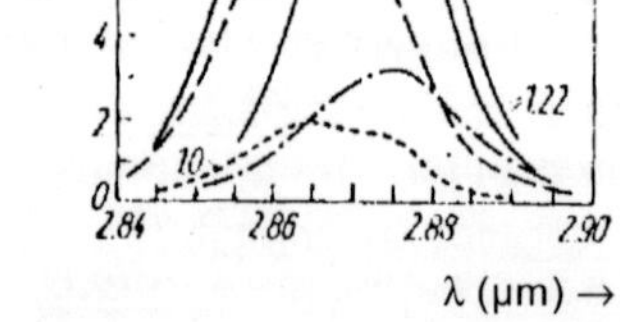

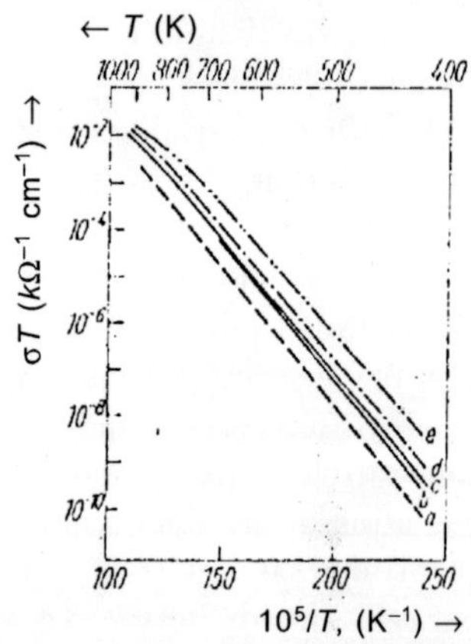

Fig.20 Effect of the Li/Nb ratio and annealing of LN crystals in the atmosphere of water vapours on the position and intensity of the absorption line of OH^--groups (wide solid, thin solid and dashed curves – annealed in H_2O vapours at 750°C; dot-and-dashed and dotted curves – after annealing (Bollmann, Stohr, 1977).

Fig.21 (right) Electrical conductivity of lithium niobate crystals in relation to the content of OH^- groups. K (2.865 μm) – a) 1.95, b) 6.8, c) 8.2, d) 17.6 cm^{-1} and e) K (2.875 μm) = 13.4 cm^{-1} (Bollmann, Stohr, 1977).

into the crystal, have a certain effect on its physical properties. It has already been shown that the effect of hydroxyl groups increases the electrical conductivity of the crystals and absorption bands appear in the optical spectrum. According to our data (Zharikov, et al. 1975), the presence of hydroxyl groups generates mechanical stresses in the crystal which lead through photoelastic constants to the variation of birefringence. It is possible that other consequences of penetration of hydroxyl groups into the crystal, undesirable from the viewpoint of practical application of the crystals, may also appear. It should be mentioned here that, from our viewpoint, the crystal does not contain hydroxyl group as such, and it contain protons which form bonds with the oxygen ions of the lattice.

As mentioned previously, annealing or growing the crystal in the atmosphere saturated with water vapours leads to the penetration of OH^- groups into the crystal lattice. This process is reversible to a certain degree. Annealing the crystal in a dried atmosphere or vacuum leads to a reduction of the number of OH^- groups but it is not possible to remove the OH^- groups completely from the crystal. It should be added here that vacuum annealing of the crystal results in a large loss of oxygen and in a reduction of niobium ions to 4^+ or even 3^+ valency state. The crystal becomes dark. The reaction of formation of OH^- groups takes place on the surface of the crystal by the reaction:

$$H_2O + O^{--} \leftrightarrow 2\ OH^- \qquad (14)$$

which is governed by the law of action of masses:

$$c(OH^-) = K_4\ (T)^{1/2}\ p(H_2O)^{1/2} \qquad (15)$$

Here $K_4\ (T)$ is the mass constant of activation of reaction (14), $p(H_2O)$ is the partial pressure of the water vapours. Equation (15) assumes that in this case the content of O^{--} ions slightly changes (oxygen loss is less than 1%).

Annealing the lithium niobate crystals in hydrogen also leads to the inclusion of OH^- ions:

$$H_2 + 2O^{--} = 2OH^- + 2e^- \qquad (16)$$

$$2Nb^{5+} + 2e^- = 2Nb^{4+} \qquad (17)$$

$$H_2 + 2O^{--} + 2Nb^{5+} = 2OH^- + 2Nb^{4+} \qquad (18)$$

The variation of the content of Nb^{5+} and O^{--} ions leads to slight reduction of the crystal. Therefore, reaction (18) is supplemented by the condition of conservation of electroneutrality:

$$c(OH^-) = x\ (Nb^{4+}) + x\ (e^-) \qquad (19)$$

where x is the molar fraction.

Usually, the crystals produced contain less lithium than the amount specified by the $LiNbO_3$ formula, because crystals with $R_m > 0.945$ are grown from a melt with the ratio $R_S < R_m$. The composition of the congruent melt ($R_m = R_S = 0.945$) remains constant during crystal growth (the congruent composition of the crystals was later determined more accurately a number of times). Therefore, the melts with the initial composition $R_m > 0.945$ or $R_m < 0.945$ are enriched or depleted in lithium, respectively, during crystal growth. In later growth stages or in growing several crystals from the same crucible, enrichment with lithium may be $R_m > 1.36$. Crystals with $R_S > 1$ grow from such a melt. It is likely that at $R_S < 1$ there are vacancies of Li^+ and O^{--} ions. At the same time, for the crystal with $R_S > 1$, the interstitial ions occupy free sites in the normal state. This means that both antiFrenkel and Frenkel disordering may take place in the lithium niobate crystals. At temperatures below the Curie point, Frenkel disordering is frozen and is therefore of no interest.

In annealing a crystal with the composition $R_S > 1$ $Li_{1+y}NbO_{3+y/2}$ in the atmosphere of water vapour, the composition $Li_{1+y}NbO_{3-z+y/2}(OH)_{2z}$ changes in accordance with reaction (14). Since the ionic radii of O^- and OH^{--} are similar, the latter may occupy the sites of both O^{--} ions and interstitial sites. A congruent crystal $Li_{0.945}$ $NbO_{2.9725}$ contains 5% of Li vacancy sites and 0.9% of sites of O^{--} ions. In annealing these crystals in water vapour, the vacancy sites of O^{--} ions are filled. These ions generate a maximum at a wavelength of $\lambda = 2.865$ μm. In annealing crystals with the composition close to congruent, maxima form at wavelengths of 2.865 and 2.875 μm. Finally, in annealing crystals with an excess of lithium, the absorption spectrum of the crystal shows a maximum at 2.875 μm.

Kovacs et al. (1990) examined absorption of OH-groups as a function of stoichiometry. The absorption band of the crystal of the congruent composition coincides with the above data. Expansion of the spectrum into components shows that the H-band consists of three lines observed previously by Fang et al. (1988). Table 3 gives the results of such an expansion of the absorption bands of the OH-groups in the Gaussian representation for crystals of different composition.

The three peaks in the absorption bands are caused by three different O–O states in the plane perpendicular to the c axis. The wave numbers and the half width of the bands are typical of weakly bonded hydrogen in the crystal lattice of lithium niobate.

Thus, analysis of the absorption spectra in the infrared region makes it possible to determine qualitatively the stoichiometry of the crystal.

7. Difference of Electrical Potentials Between the Crystal and the Melt

In their work, D'yakov et al. (D'yakov, Shumov, Rashkovich, Aleksand-rovskii, 1985) recorded the emf formed during solidification of lithium niobate from the melt at a temperature of 1520 K at which the melt and the crystal are ionic

Table 3 Wave numbers, halfwidths of maxima at half height and absorption factors of vibrations of OH groups in lithium niobate crystals with different ratios Li/Nb. $(Li/Nb)_m$ is the molar ratio for the melt, $(Li/Nb)_c$ for the crystal

$(Li/Nb)_m$	$(Li/Nb)_c$	Wave number cm^{-1}	Half width cm^{-1}	k, cm^{-1}
0.818	0.899	3487	31	2.08
		3467	12	0.14
0.945	0.945	3485	28	2.43
		3468	11	0.60
1.000	0.959	3493	16	1.63
		3482	13	1.89
		3470	11	1.34
1.200	0.989	3482	10	0.86
		3481	10	2.12
		3467	9	3.21

conductors. Under the conditions close to equilibrium, in addition to the electrical phenomena, associated with the thermal emf of the melt and the crystal, there was also an emf proportional to the solidification rate. This emf may be caused by neutralization of the charge, trapped by the growing crystal, as a result of electrical conductivity of the medium.

Emf was measured using two platinum–rhodium thermocouples which were also used as electrodes. One of the thermocouples was immersed in the melt, the other one rotated in the crystal. Figure 22 shows the dependence of the potential difference φ between the two thermocouples on temperature in the crystal T_1 at a constant temperature of the melt T_2. Point 1 corresponds to the start of drawing the crystal where both thermocouples were in the melt. Point 2 corresponds to the moment when the junction of the thermocouple is situated in the vicinity of the solidification front. Point 3 corresponds to the constant drawing rate, point 4 to an arrest in drawing, 5 to the start of lowering the crystal, 7 to melting of the thermocouple from the crystal. The resultant dependence indicates the contribution to the value of φ of the thermal emf of the melt (thermoelectric coefficient α_l) and of the crystal (α_2) and of the emf proportional to the growth rate v. In section 1–2 where both thermocouples are in the melt, the emf is determined by the expression:

$$\varphi = \alpha_l (T_1 - T_2) \tag{20}$$

Quantity α_l, determined from two independent experiments is equal to $\alpha_l = -0.4$ mV K^{-1}. In sections, 3–6 when the thermocouple is in the crystal, the following equation holds:

$$\varphi = \alpha_l (T_1 - T_k) + \alpha_S (T_k - T_2) + \alpha_V V \tag{21}$$

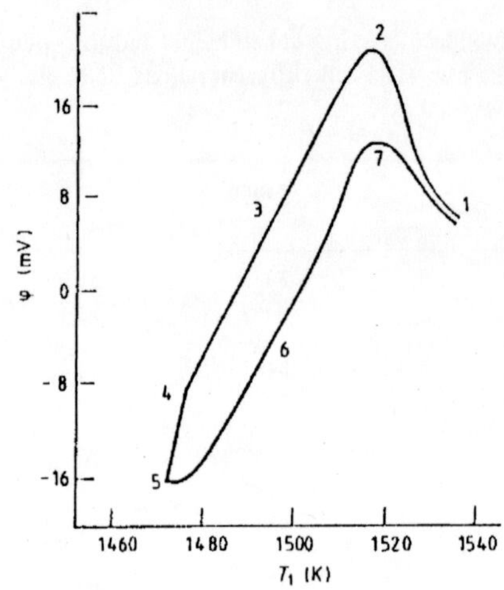

Fig.22 Emf of the crystal–melt system φ as a function the temperature of the electrode inserted in the crystal. The numbers of the curve refer to growth stages mentioned in the text (D'yakov, et al., 1985).

where T_k is the melting point, α_V is the coefficient of solidification emf. In section 2, the melt temperature T_2 and the growth rate are maintained constant, and the slope of the curve is determined as $d\varphi/dT = \alpha_S$. The results of more than 30 experiments were used to determine the mean value $\alpha_S = (0.76 \pm 0.02)$ mV K^{-1}. Different signs at α_S and α_1 result in the appearance of a maximum on the $\varphi(T)$ curve in the vicinity of point 2. α_V, determined from the hysteresis loops, is $\alpha_V = (1.25 \pm 0.2)$mV s μm^{-1}.

The values of α_V and α_S can be used to control the solidification rate and determine the sudden increase of the growth rate. The appearance of crystallization emf, proportional to the growth rate, is obviously associated with perturbation of the double electric layer existing at the phase boundary. The observed phenomenon may be typical of a large number of oxide crystals.

8. ResidualThermal Stresses in Crystals

The relaxation of thermal stresses in the crystal takes place by means of plastic flow of the material as a result of movement of dislocations and diffusion of point defects up to the moment of exit of the crystal from the plasticity zone during growth. In cooling below the plastic deformation temperature, the regions of the crystal which prior to plastic deformation were compressed and, consequently, lost part of the material, become stretched when a lower temperature was reached and vice versa. Thus, the cooling crystal contains residual stresses whose magnitude is similar and the sign is opposite of that of thermal stresses. The residual stresses are determined by the axial and radial temperature gradients. If the curvature of the growth surface is small (flat front), the controlling role is played by the axial distribution of temperature and the radial gradient may be ignored.

An approximate evaluation of the thermoelastic stresses in the growing crystal may be carried out using the equation proposed by Chernov (Chernov, 1980)

$$\sigma_{max} = (K/8)\, \alpha\, E\, T''\, D^2 \tag{22}$$

where σ is the component of the stress tensor, K is a parameter, α is the linear coefficient of thermal expansion, E is Young's modulus, D is the size of the specimen in the direction of temperature variation, T'' is the curvature of the temperature curve. The above equation shows (this is suitable for practice) that it is more efficient to anneal crystals starting at the growth temperature, without preliminary cooling to room temperature.

8.1 Thermal stresses formed during growth

It is not possible to draw a crystal in the absence of an axial temperature gradient, if the crystal length is considerably greater than is diameter. The radial gradient should also form with a corresponding thermal stress. To evaluate the critical value of the temperature gradients at which there is still no cracking of the crystals during growth, Brice (Brice 1977) carried out an approximate analysis developed in (Brice, 1968). Analysis assumes the constant temperature of environment T_0 and also that part of the crystal surface at temperature T loses the heat into the environment at the rate $\sigma(T - T_0)$. If melting point is T_m, the distribution of temperature at the point (r, z) in a long crystal with radius R is described by the equation:

$$T = T_0 + (T_m - T_0)\frac{(1 - hr^2/2R)}{1 - \frac{1}{2}hR}\exp\left[-\left(\frac{2h}{R}\right)^{1/2} z\right] \tag{23}$$

where h is the cooling constant (ratio of σ to thermal conductivity), the origin of the coordinates is at the axis of rotation on the growth surface. Equation (23) shows that at $r = R$ we can write

$$\mathrm{d}T/\mathrm{d}r = (hR/2)^{1/2}\frac{\mathrm{d}T}{\mathrm{d}z}\left(1 - \frac{1}{2}hR\right)^{-1} \tag{24}$$

Equation (24) shows that the mean radial temperature gradient for constant z is equal to half the value at $r = R$. Thus, the temperature difference along the radius is determined by the equation:

$$\Delta T_R = R/2(hR/2)^{1/2}\mathrm{d}T/\mathrm{d}z(1 - hR/2)^{-1} \tag{25}$$

The equations derived by Brice show that the maximum tensile stress in the axial and tangential directions forms at $r = R$ and in every individual case reaches the value $1/2(\alpha\,\Delta T_R)$. Consequently, the resultant stress is

$$\varepsilon_{max} = \alpha\,\Delta T_R\,/\,2^{1/2} \tag{26}$$

where α is the expansion factor. To grow a crystal without cracking, it is essential to fulfil the inequality $\varepsilon_{max} < \varepsilon_b$. The maximum acceptable temperature gradient is calculated from the equation

$$(\mathrm{d}T\,/\,\mathrm{d}z)_{max} = (4\,\varepsilon_b\,/\,\alpha\,R^{3/2})[1\,/\,h]^{1/2}\,(1 - h\,R\,/\,2) \tag{27}$$

Table 4 gives the measured values of the corresponding parameters for lithium niobate crystals. These data give the maximum temperature gradient at which the crystal remains in one piece, $(\mathrm{d}T/\mathrm{d}z)_{max} = 78/R^{3/2}\,°\mathrm{C\ cm^{-1}}$, for small crystals. The limited number of data available for lithium niobate are in agreement with the calculated gradient. It should be mentioned that the above equations for calculating the temperature gradients were tested only on crystals of small diameter. Approximation to crystals of larger diameter requires special examination. For the crystals with $R = 1$ cm, the maximum gradient is around 50°C cm^{-1}.

Table 4 Properties of crystalline lithium niobate

Thermal expansion	16×10^{-6} (a), 2×10^{-6} (c)
Heat capacity (cal cm^{-3})	0.8
Thermal conductivity (cal cm^{-1} s^{-1} K^{-1})	2×10^{-3} (mpt) 1×10^{-2} (300 K)
Cracking stress	2×10^{-4}
Cooling constant (cm^{-1})	0.5

8.2 Thermal stresses formed in cooling

All crystals are annealed after growing and can be heated and cooled in subsequent stages in, for example, securing for cutting. We shall examine a long rod with radius R, thermal conductivity K and heat capacity per unit volume c. It is assumed that the cooling rate remains constant (the conditions do not change) and, consequently, the energy conservation condition is fulfilled:

$$\pi r^2\,c\,\mathrm{d}T\,/\,\mathrm{d}t = 2\,\pi\,r\,K\mathrm{d}T\,/\,\mathrm{d}r. \tag{28}$$

We introduce the notation $k = K/c$ and transform equation (28)

$$dT / dt = (2k / r)dT / dr. \tag{29}$$

From (29), we obtain the radial temperature difference

$$\Delta T_R = \left(R^2 / 4k\right) dT / dt \tag{30}$$

Using equation (26) and the condition $\varepsilon_{max} < \varepsilon_b$, it is possible to calculate the maximum cooling rate at which the crystal does not crack

$$[dT / dt]_{max} = 8^{1/2}\, k\varepsilon_b / \alpha R^2. \tag{31}$$

Taking into account the above equations of the constants, presented in Table 4, we calculate all required quantities. Defining the crystal radius $R = 3$ cm and the axial gradient $dT/dz = 10°C\ cm^{-1}$, the radial temperature gradient is $\Delta T_R \simeq 52°\ cm^{-1}$. We calculate the maximum stress from equation (26) $\varepsilon_{max} = 5.9\times10^{-4}$, specify the permissible stress $\varepsilon_b = 3\times10^{-4}$ and calculate the permissible axial temperature gradient using equation (27) $(dT/dz)_{max} = 40°C$. It should again be stressed that all these calculations are only of the highly approximate nature.

Carruthers et al (Carruthers, Peterson, Grasso, Bridenbaugh, 1971) calculated the mutually dependent temperature gradients and growth rates of lithium niobate crystals with different lithium oxide content. Unfortunately, these calculations were carried out for the compositions in which the lithium oxide content was higher than 50% and did not include crystals with a congruent composition which are of special interest. The results of the calculations are presented in Fig.23.

The cracking susceptibility of crystals during cooling after growth can be examined from the viewpoint of the temperature dependence of crystal lattice distortion. The temperature dependence of lattice distortion for crystals with stoichiometric and congruent compositions was examined in detail by Taylor (Taylor, 1990) and is shown in Fig.24. The graph shows that the most extensive deformation takes place in the range 300–500K. We believe that in this temperature range it is necessary to reduce the temperature very smoothly and this is often not considered in the technology of production of lithium niobate crystals.

9. Determination of the Chemical Composition of the LN Crystal

The inspection and determination of the chemical composition of the melt and of a crystal grown from it is an important compound part of technology of high-quality crystals. As already mentioned several times, a congruent melt is required for producing crystals with a uniform chemical composition. The con-

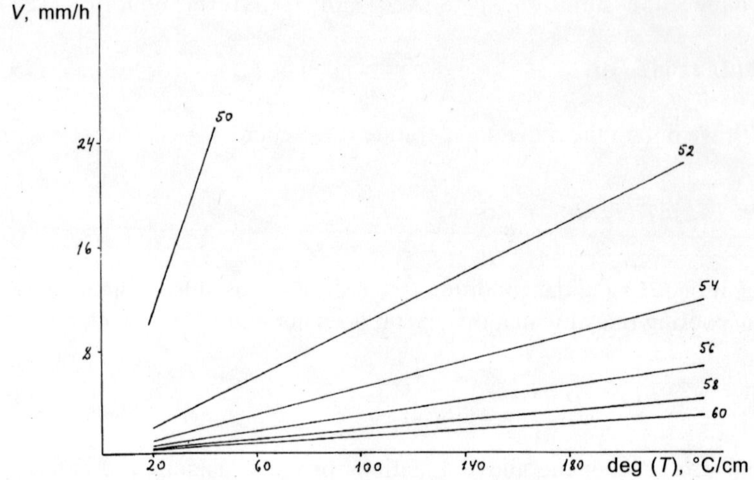

Fig.23 Minimum axial temperature gradient in the melt *G* required for preventing concentration of super-cooling of melts of various composition and different pulling rates (Carruthers, Peterson, et al, 1971).

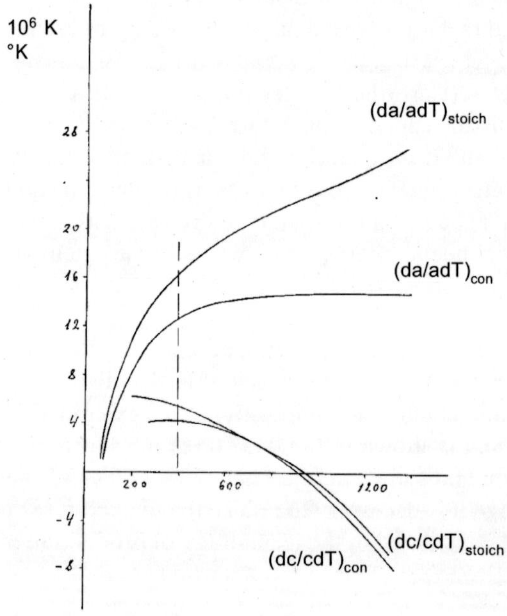

Fig.24 Instantaneous expansion coefficients (10^6/K) for stoichiometric and congruently melting lithium niobate. The vertical line determines the zone with high probability of cracking of the crystal during its cooling (Taylor, 1990).

gruent composition is determined mainly by a shortage of oxygen in the initial niobium pentoxide, the impurity composition of reagents used and also by technological parameters. Therefore, the congruent composition differs for different reagents used and must be determined again if a new batch of the starting material is used.

As a result of a large permissible error, the classic chemical methods of analysis are not suitable for this purpose. There are a large number of physical methods which enable the chemical composition of the crystal to be determined with various degrees of reliability. We shall discuss some of them.

9.1 Determination of the Li/Nb ratio from Curie Temperature T_C

The technical aspect of the problem has been discussed in considerable detail by Fenske and Kuz'minov, 1988. The dependences of the ferroelectric phase transition, characterised by the Curie temperature in relation to the content of the molar fraction of Li_2O in a crystal, are represented in Fig.25. The measurement results cover the range from 47.0 to 49.0 mol.% Li_2O in the initial melt which corresponds to the majority of specimens in the composition range 48.0–49.0 mol.%. The results processed by the method of least squares yielded a linear dependence

$$T_C = -637.30 + 36.70\, c \qquad (32)$$

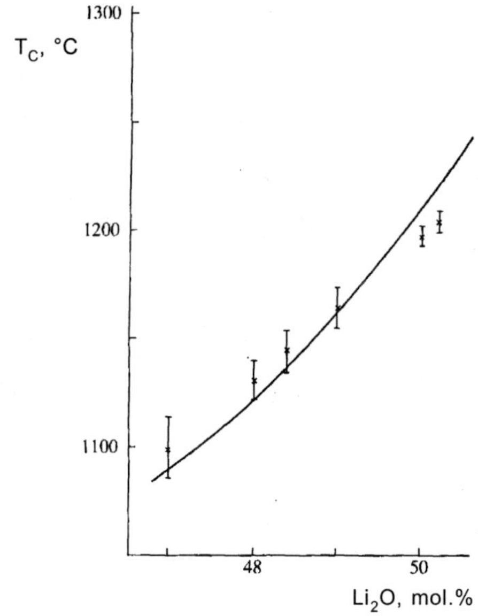

Fig.25 Dependence of the Curie temperature of the Li_2O content of a lithium niobate crystal. The curve was constructed on the basis of equation (34). Experimental points were taken from Brice, 1977.

or

$$c = 17.37 + 0.02725\ T_C, \qquad (33)$$

where T_C is the Curie temperature in °C, and c is the Li_2O content (mol.%) in this specimen. The Curie temperature is determined more accurately by the quadratic dependence

$$T_C = 9095.2 - 369.05\ c + 4.228\ c^2 \qquad (34)$$

(the curve is shown in Fig.25). The Curie temperature was measured highly accurately in a study by Bordui, Norwood, et al, 1991. T_C was measured using the standard method of dielectric measurements (Turner, Nash, Bridenbaugh, 1970). Temperature was varied periodically and the ferroelectric phase transition was recorded in the form of a sharp peak. The measured values of T_C did not depend on frequency in the range from 10 kHz to 1 MHz, the signal amplitude was 1 V at a frequency of 300 kHz. The furnace temperature was varied at a rate of 1°C min^{-1}. The furnace was heated from room temperature to T_C for 90 min, the thermal cycles were approximately 10°C from peak to peak. There was no thermal hysteresis as a result of placing the examined specimen into a gradientless zone in the furnace. The absolute accuracy of measuring T_C was $\pm$ 0.3°C, reproducibility of the measurements of T_C was $\pm$ 0.1°C. This temperature reproducibility was obtained by computer control of the heating of the furnace.

It is quite difficult to determine directly and inspect the numerical value of the composition of the Li/Nb melt used in crystal growth. The weight loss of the melt due to evaporation of lithium was determined by a special procedure. The loss was less than 5×10^{-6} g/g of melt/day. Summing up all possible sources of errors, the absolute error of determination of the Li_2O content in a steady melt was $\pm$0.01 mol.% Li_2O. The numerical value of the Li_2O content in the melt was initially determined by growing a crystal boule by the Czochralski method along the Z axis with a diameter of approximately 80 mm and 150 mm long from various compositions of the starting melt. The grown crystals contained approximately 85% of the solidified melt. Curie temperature T_C and phase matching temperature T_m were measured on specimens from the beginning and end of each boule. Measurements showed that the congruent composition is between 48.35 and 48.40 mol.% Li_2O and the Curie temperature T_C = 1142°C. Special attention was paid to the method of determining the congruent content because this is a very important parameter of crystal technology, and difficulties in determining this parameter were also shown.

9.2 Determination of the Li/Nb ratio from phase synchronism temperature T_m

The variation of the phase matching temperature T_m can be associated with the composition of the melt R = [Li/Nb]. For phase matching of the primary light wave normal to the optical axis of the crystal (the so-called non-critical synchronism), with the excited wave of the second harmonics the birefringence of the crystal for this wave length $B = n^o_2 - n^e_2$ should be equal to the dispersion of the crystal $D = n^o_2 - n^e_1$, i.e. $B - D = 0$. The variation of the value of B–D in relation to the composition of the crystal R should be compensated by the variation of phase matching temperature T_m. This condition can be written in the form:

$$[d(B - D)/\mathrm{d}T]\ \Delta T = [\mathrm{d}(B - D)/\mathrm{d}T]\ \Delta R \qquad (35)$$

It is well know that the birefringence of NL depends on the melt composition and is controlled by the extraordinary refractive index n^e. The graph shown in Fig.26 can be used to determine the angle of inclination of the straight line in the vicinity of the composition $R = 1$ (for $\Delta R = 0.1$), $\Delta B = 0.0072$ and T_m = 122°C $d(B - D)/\mathrm{d}t = -5.9\times10^{-5}\ \mathrm{K}^{-1}$. Thus, knowing the birefringence, we can calculate the phase synchronism temperature and determine the chemical composition of the crystal. The linear dependence $T_m = f(R)$ is disrupted at $R > 1.1$, probably as a result of the increase of $\mathrm{d}D/\mathrm{d}R$, and saturation R in the crystal with increasing lithium content of the melt (Turner et al, 1970). The phase synchronism temperature as a function of the molar fraction of Li_2O in the liquid phase is shown in Fig.26. The dependence is approximately linear and increases with increasing Li_2O content of the melt. A linear dependence was obtained by processing the experimental data by the method of least squares and can be described by:

$$T_m = 50.5 + 32 \times 10^2\ (x - 0.50) \qquad (36)$$

where x is the molar composition of the crystal. The standard deviation from the straight line is $\pm$7°C.

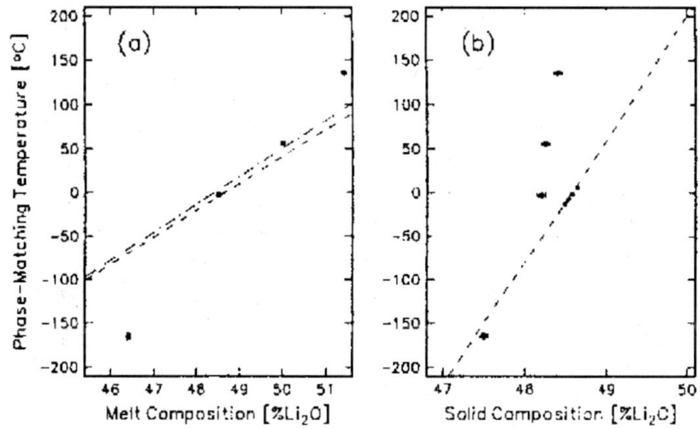

Fig.26 Dependence of the temperature for noncritically phase-matched SHG in $LiNbO_3$ vs (a) melt and (b) solid composition. Squares: Schmidt et al, 1991; circles: Chow et al, 1991; dashed line Noda et al,; dash-dotted line: Fay et all.

In a study by Bordui et al, 1991, T_m was determined using a Nd:YAG laser with 1.064 μm. The yield of the second harmonics λ = 0.53 μm was recorded with a silicon detector and a power measuring device. The linear dimensions of the specimens for examination were 4×10×4 mm, along the radiation direction y = 10 mm. The y-surfaces were processed by a procedure typical of optical specification. The accuracy of orientation of the y-direction was 30'. Initially, T_m was determined on non-polarised specimens and this resulted in considerable broadening of the maximum of the yield of the second harmonics. According to the authors, this broadening and even its disappearance was caused by the fact that in the vicinity of the congruent point on the side of a shortage of Li_2O the crystal contains mainly domains of the same sign, and in transition of the congruent point to the region of a surplus of Li_2O there are mainly domains of a different sign. To eliminate these effects, it is recommended to measure the T_m of monodomain specimens. The domain structure of lithium niobate crystals and the mechanism of formation were studied in detail by the author in books (Kuz'minov, 1987; Prokhorov, Kuz'minov, 1990). We shall now discuss the problems of observing the domain structures in lithium niobate crystals for identification of the domain state of the crystals since the polydomain crystals cannot be used in physical devices.

9.3 Determination of the composition of a crystal from the speed of propagation of a surface of acoustic wave (SAW)

To determine the chemical composition of lithium niobate and lithium tantalate crystals, the authors (Kushibiki, Takahashi, Kabayashi, Chubachi, 1991) proposed to utilise their elastic properties, especially the speed of propagation and damping of surface acoustic waves (SAW). This method is used widely in both scientific investigations and in industrial production of crystal sheets of lithium niobate for devices based on SAW. The above authors carried out their investigations on 128° sheets of the XY cut, 75 mm in diameter. The relative accuracy of measuring the speed was ±0.005% at a point and ± 0.02% when scanning a space of 75×75 mm. Calibration was carried out on specimens cut out from crystal grown from different starting melts. The composition of the crystal was determined from the Curie temperature, as described in section 9.1. Crystal density was also determined and examined as the mean value for sheets in the radial direction and in the sheet thickness. Figure 27 shows the results obtained in measuring the speed, density and chemical composition of a relatively large batch of 128° sheets of the YX cut. The graph shows clearly that the speed of the SAW and the crystal density depends strongly on the crystal composition. When the Li_2O content of the crystal is increased the speed of the SAW increases linearly to 37.16 m/s/mol.%, and in the vicinity of the congruent composition the density decreases to 7.25 kg/m^3/mol.%. The sensitivity of determination of the composition was 0.02691mol.% m/s. A resolution power higher than 0.005 mol.% corresponds to a 0.2°C error of determination of the Curie temperature. The graph shows that congruent composition crystals give sheets of high quality with homogeneous elastic properties. The congruent com-

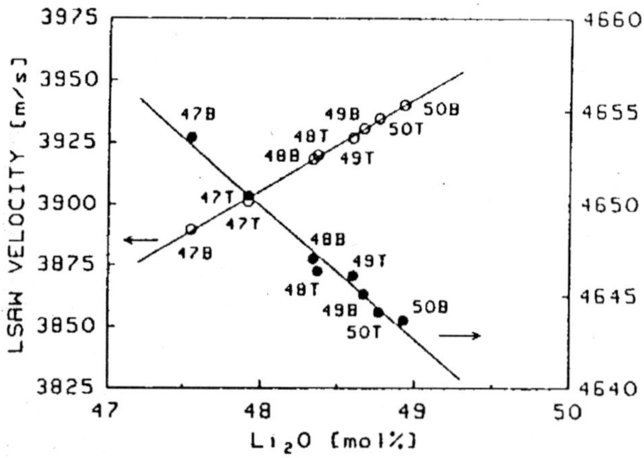

Fig.27 Experimental dependence between the rates of SAW, melt composition (mol.% Li_2O) and the density of sheets of 128° *yx*-cut of lithium niobate crystals (Kushibiki et al, 1991).

position of the crystals was determined in the vicinity of 48.45 Li_2O mol.%, the density of the crystals with congruent composition was 4647.4 kg/m^3. The reproducibility of the results was verified under the conditions close to those in industrial production of the sheets. There were small deviations in the rate of propagation of the SAW and were caused, according to the authors, by small changes of the chemical and thermal conditions in the melt during crystal pulling. The mean speed of the SAW of the congruent composition crystals was 3922.07 m/s, the maximum deviation for the mean value was 3.6 m/s. The 'effective' congruent composition of the melt, determined in this work, is similar to that determined by O'Bryan (O'Bryan, Gallagher, Brandle, 1985).

As already mentioned, the use of a congruent melt for growing crystals greatly increases the yield of suitable devices based on SAW and stabilises their physical properties. As an example, Fig.28 shows the histogram of distribution of filters with respect to absorption. The filters produced from congruent composition crystals have a narrow absorption band (Bordui et al, 1991).

To simplify the determination of the composition of the crystal from which sheets were produced for devices based on SAW, the author of this book, us-

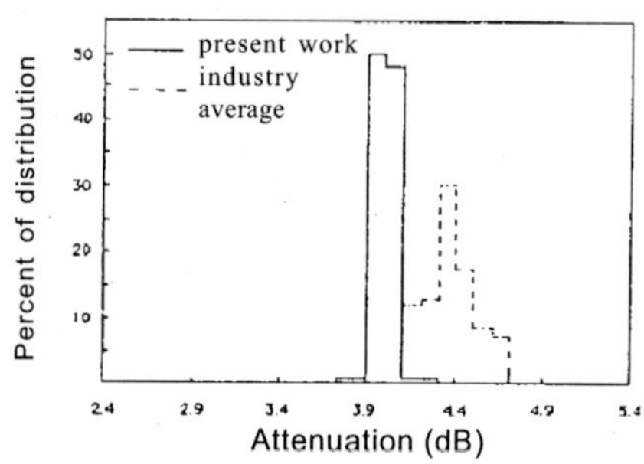

Fig.28 Histogram of the losses of SAW filters based on lithium niobate crystals in relation to the composition of the melt (Bordui et al, 1991).

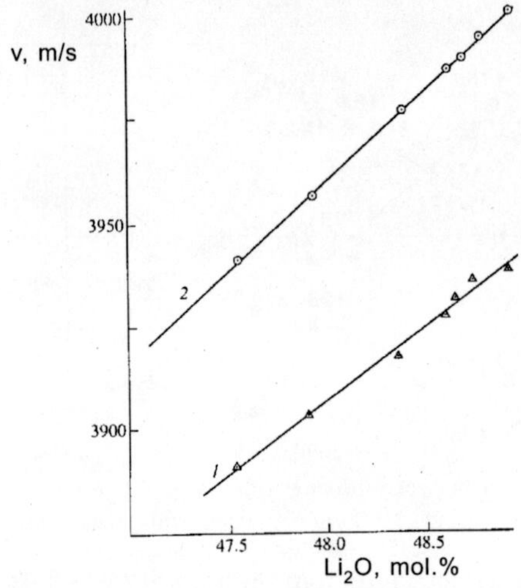

Fig.29 Experimental dependences of the speed of SAW and crystal composition: (1) Kushibiki et al, 1991; (2) Yamada et al, 1987. Sheets of 128° YX cut of lithium niobate crystal.

ing the results published by Kushibiki et al, 1991 (1) and Yamada et al, 1987 (2), plotted the dependence of the speed of SAW on 128° sheets of the YX orientation on the composition. The graph plotted using a single scale is shown in Fig.29. However, it should be mentioned that the speeds of SAWs, given the above studies, greatly differed for sheets with the same orientation. It is possible that the authors considered different compositions – the composition of the melt and the composition of the crystal grown from this melt. It is well known that the composition of the melt coincides with that of the crystal only at a congruent point but there is no such correspondence even for the congruent composition. These problems are discussed in Section 20. In this method of determining the chemical composition of the crystal errors may occur as a result of the inaccurate determination of the crystallographic orientation on the sheet. Figure 30 shows the dependence of the speed of the SAW on the orientation of the sheet in the vicinity of the selected 128° YX-orientation. As a result of slight rotation, the sheet is set to the maximum speed and the measurements are taken out along this direction. The sheet orientation can be determined more accurately in an x-ray meter.

Cutting a whole crystal boule into sheets and measuring subsequently the SAW speed, it is possible to plot the curve of distribution of the impurity whose role is played by Nb, if the melt composition differs from congruent.

10. Formation and Breakdown of Lithium Triniobate $LiNb_3O_8$ in the Volume and on the Surface of Lithium Niobate Crystals

Lithium niobate crystals with congruent composition are metastable at temperatures

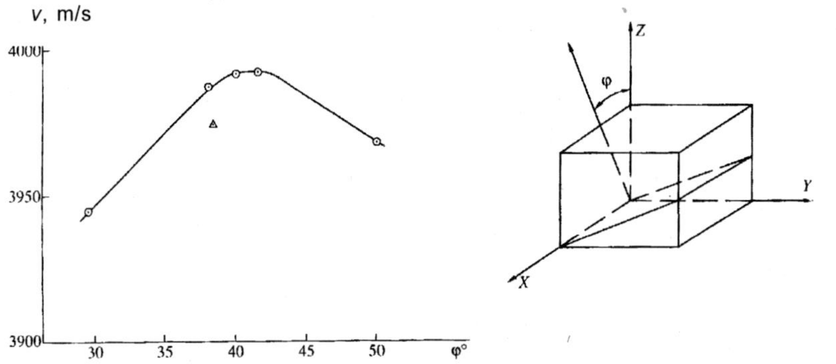

Fig.30 Dependence of the speed of SAW along the *X* axis in a sheet of a lithium niobate crystal 128° *yx*-cut on angle ϕ of the deviation of the direction of the normal in relation to the *Z* axis plotted using the data by Hinkov, 1990. Right: orientation of the sheet in the *X, Y, Z* coordinate system.

below 900°C as a result of special features of the Li_2O–Nb_2O_5 phase diagram (Prokhorov, Kuz'minov, 1990). The crystal sheets are heat treated in polarization and manufacture of elements for integrated optics and acoustoelectronics. Experiments show (Arminise, Canali, et al, 1983) that the phase composition of the surface changes: lithium triniobate $LiNb_3O_8$ (LTN) forms in the temperature range 300–700°C in the subsurface layer. The second phase also forms in the volume of the crystal, as indicated by the appearance of opalescence in long-time annealing a crystal with a congruent composition. Formation of LTN on polished sheets makes it possible to examine the thermal conditions of this process. With a further increase of temperature the amount of the second phase decreases and LTN disappears at 900°C, in accordance with the equilibrium diagram.

Nassau, Levinstein, Loiacano, 1966 showed that mechanical polishing of sheets is accompanied by the formation of an amorphous layer 50–300 μm thick. Under the effect of temperature mass transfer processes are activated in this layer and change the morphology, structure and composition of the surface. Experiments have shown that the surface roughness already increases at $T = 300$°C and individual faceted hillocks appear. Analysis of electron diffraction patterns shows that a two-phase system: $LiNbO_3$ and $LiNb_3O_8$, forms in the subsurface layer. In bulky lithium niobate crystals the second phase precipitates only after long-time annealing at temperatures above 600°C (Svaasand et al, 1974). A reduction of the temperature and duration of formation of LTN on the surface of the polished sheet is explained by a higher diffusion rate of cations in the amorphous layer in comparison with the crystal.

Formation of $LiNb_3O_8$ from $LiNbO_3$ requires a surplus of niobium on the surface of the sheet which may form as a result of growth of crystals from a congruent melt containing a surplus of niobium.

When annealing temperature is increased to at least 600°C the rate of formation of the second phase is high. A matt deposit forms on the surface as a result of the formation of oriented islands of LTN 100–150 Å and 0.15–0.5 μm in size, with a density of 10^7–10^{-10} cm^2. Usually, these islands group in the vicinity of scratches. The optimum temperature range for the liquid-phase epitaxial synthesis of LTN on the basal plane of the LN is 750±50°C.

Faceted islands of LTN are oriented in relation to the LN substrate as follows: $(\bar{3}01)[010]_{ph} \| (0001)\langle 2\bar{1}\bar{1}0 \rangle_{mat}$. The forms of growth of LTN are represented mainly by pinacoids $(\bar{h}00)$ and $\{h00\}$, stretched along the [010] direction.

The mutual orientation of the matrix and the phase will now be examined. LN and LTN crystals belong to different structural types: LN belongs to the spatial group $R3c$ with lattice parameters $a = 5.15$ Å and $c = 13.86$ Å (Abrahams, Reddy, Bernstein, 1966); LTN crystallises in a monoclinic crystal system (spatial group $P2_1/a$) with parameters $a = 5.033$ Å, $c = 7.457$ Å and $\beta = 107.34°$ (Lundberg, 1971). In accordance with the point group symmetry $3m$ the lithium niobate crystals contain three equivalent directions $<11\bar{2}0>$coinciding with the directions of the hexagonal axes x, y, u, and three planes of symmetry of the type $\{11\bar{2}0\}$ perpendicular to the axis. The microcrystals of LTN are oriented on the surface of the LN in such a manner that each of the three planes $\{11\bar{2}0\}$ of the LN coincides with the (010) plane in the LTN crystal, with the unique plane of symmetry in the point group $2/m$ to which the LTN belongs. Therefore, the LTN microcrystals occupy three equivalent positions on the substrate rotated by 120° and form dendrite joints under angles of 60 and 130° (Semiletov, Bocharova, Rakova, 1987). Li and Nb atoms of the same type form chains in the coinciding directions $[\bar{2}110]$ and [010] and the mismatch of the interatomic spacing of these chains is 2.27%. When a new phase forms, the structure of LN is rearranged in such a manner that the oxygen frame is only slightly distorted. Filling of octahedral cavities with cations in LTN takes place with the chains of the metals of the same type in the plane of intergrowth of LTN and LN remaining parallel.

When the annealing time is increased, the LTN islands grow, merge together, and form a continuous layer. The rate of formation of the LTN is governed by the dependence $v{\sim}t^{1/2}$ and is limited by the diffusion process.

The monoclinic phase starts to break down at temperatures above 800°C: the faces become round and their size decreases. In the temperature range 800–850°C a dynamic equilibrium forms between the formation and breakdown of the monoclinic phase. Annealing at temperatures above 900°C leads to complete dissociation of LTN islands which leave depressions with steps spreading from them. The time required for complete disappearance of a LTN layer 150–200Å thick at $T = 950°$C is around 2 hours. The density, dimensions and morphology of the LTN islands depend greatly on the quality of surface treatment: as the quality increases the density of the LTN islands decreases. Semiletov et al (Semiletov, Bochova, Rakova, 1987) link the distribution of the islands of the monoclinic phase on the surface of an LN sheet with the disruption of stoichiometry in the crystal volume. Thus, the precipitated second phase may decorate heterogeneities of the crystal composition. In all likelihood, this concerns the distribution of the islands of the monoclinic phase in the specimens subjected to recrystallisation annealing. In polydomain crystals, lithium triniobate island are distributed along the domain boundaries. It may therefore be assumed that the regions of LN crystals in the vicinity of the domain boundaries are enriched with niobium, as confirmed by the results published by Mileshina, Belugina, et al 1984.

In lithium niobate crystal technology, especially crystals of congruent composition, it is important to take into account the possibility of formation of a monoclinic phase in the crystal volume during long-term annealing in the process of polarization. It is important to select the optimum annealing conditions not causing precipitation of the second phase but preventing crystal cracking. Figure 31 shows the variation of the morphology of the basal face in relation to the annealing time of the crystal sheet.

11. Macromorphology of LN Crystals Grown by the Czochralski Method

The external form of a crystal is determined by the growth conditions and the anisotropy of growth rates along crystal directions. In the Czochralski method, the crystal is rotated and is pulled. The crucible with the melt and the hf induction coil are axial with respect to the axis of rotation of the crystal. Therefore, if the symmetry of the thermal field is not disrupted, the crystal has the form of a body of revolution and belongs to the point symmetry group m/∞. The crystal is cylindrical if grown along the axis of symmetry C_3, and in growth along the x or y axis the crystal has a flattened form. Relative speeds of growth along crystallographic direction [*hkl*] were determined for the LN crystal:

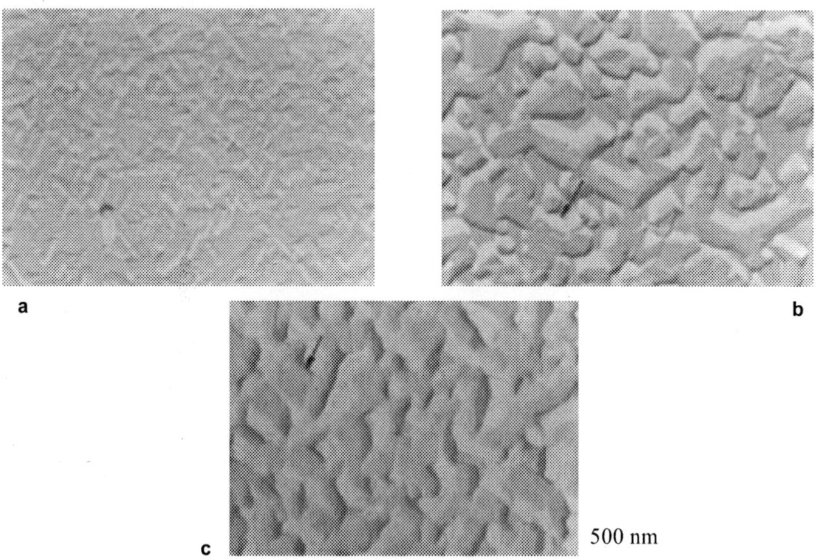

Fig.31 Variation of the morphology of the basal face of a lithium niobate crystal in relation to annealing time: a) 1 h ; b) 12 h; c) 20 h ($T = 750°C$) (Bocharova, 1987).

$V_{100} = V_{120} > V_{001}$.

A typical feature of LN crystals is the formation of grooves on the lateral surface of the boule. The configuration of the groove depends on the growth direction. The distribution of the grooves on the crystals with different pulling directions is shown in Fig.32 in perspective and in the cross-section. The number of grooves on the cylindrical part of the crystal differs from that of the grooves formed on the conical or seed part of the crystal. When a steady growth condition is reached, some of the grooves disappear. Figure 33 shows schematically the vertical sections of grooves which indicate that they are the result of periodic growth of certain planes during rotation and pulling of the crystal. These planes are normal to the mirror planes of a crystal drawn in the direction of the *c* axis and are determined by the ($h01$) indices.

Nizeki, Yamada and Toyoda (1967) explained the formation of ridges on the basis of a proposal for the growth mechanism of a LN single crystal from a melt according to which the ($h01$) planes have the highest tangential speed of growth of all other planes. The proposed mechanism of gradual formation of the grooves is shown in Fig.33. In this graph, vector $\bar{n}$ is the normal to the ($h01$) plane parallel to the corresponding vector of the lattice shown in Fig.33e. The growth vector $\overline{G}$ is situated in the vertical plane together with the pulling direction *P* and is parallel to the lateral surface of the crystal which is in contact with the melt surface. Vector $\overline{G}$ coincides with growth direction *P* when the stable growth condition is reached. In the initial stage of growth with the gradual increase of the radial dimensions of the crystals, the crystal forms a

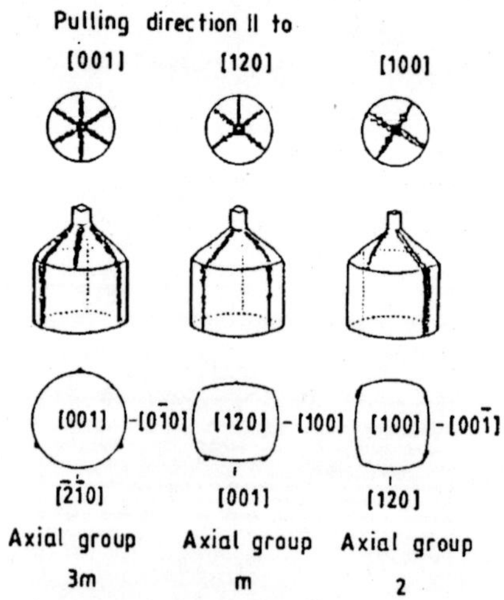

Fig.32 Configuration of growth ridges on the surface of a lithium niobate crystal. Top row: ridges on the shoulder viewed from the seed. Middle row: perpsective view. Bottom row: cross-section normal to the pulling direction (Niizeki, Yamada, Toyoda, 1967).

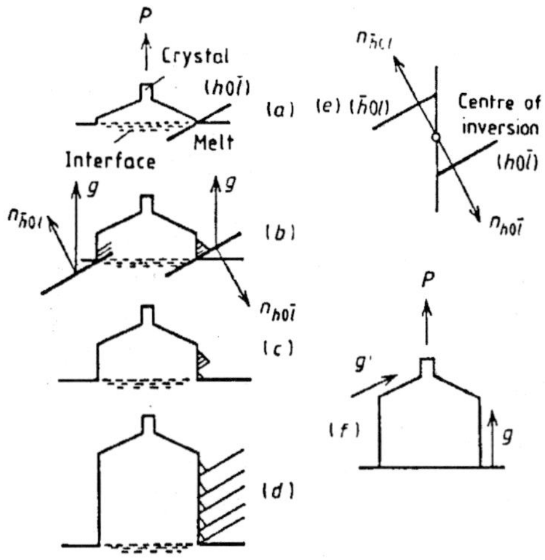

Fig.33 Mechanism of gradual formation of growth ridges. Explanation is given in the text (Niizeki et al, 1967).

sharp angle with the melt surface (Fig.33f). Two cases in relation to the position of two vectors $\overline{n}$ and $\overline{G}$ will be examined. The first case is the one in which the angle between the vectors $\overline{n}$ and $\overline{G}$ is obtuse. Nucleated on the surface, shown in Fig.33a, the plane $(h0\overline{1})$ grows very rapidly. If the crystal is continuously pulled out of the crucible, a small amount of the melt will be retained on the growth surface (Fig.33b) until the balance between the viscosity, surface tension and gravitation is disrupted. In this case, the surface of the melt is lowered to the initial level and the remnants of the melt are displaced to the crystal surface (Fig.33c). This process is repeated during pulling so that grooves, shown in Fig.33d, form on the crystal surface. In the second case, the angle between $\overline{n}$ and $\overline{G}$ is acute. In this case, the $(h0\overline{1})$ planes will always grow into the bulk of the melt and no ridges form on the crystal surface (Fig.33b).

As mentioned previously, the direction of vector $\overline{G}$ changes during crystal growth from initial to a later stable state. Some vectors $\overline{n}_{hkl}$ can form obtuse angles with $\overline{G}_0$ in the initial stage which then change in the final stage. These planes form grooves only on the expanding part of the crystal.

The planes responsible for the formation of ridges were indexed on the basis of trigonal pyramids which form in large quantities in the lower part of the crystal and their planes are parallel to the planes in the grooves. The angles between the C axis and the direction normal to the planes of the pyramids equal 57° which corresponds to the normal to the $\{10\overline{2}\}$ planes in the hexagonal co-ordinates system and to the axis $\{110\}$ in the rhombic system. The system of planes forms a rhombohedron with an obtuse angle at a tip of 99° 16'.

When the power of the rf generator is switched off, the melt in the crucible solidifies with the formation of square crystals on the surface. X-ray examination of these crystals confirmed that they belong to the rhombohedron planes $\{10\bar{2}\}$. Experiments also show that the $\{10\bar{2}\}$ planes can grow very easily in the melt being cooled.

The Curie principle shows that elements of the symmetry of crystals form under the effect of external influences like elements typical of the crystal without any effect and symmetry of action. The external effect in crystal growth by the Czochralski method can be regarded as a polar axial group of symmetry $C_{\infty V}$. In this case, crystal symmetry without an effect corresponds to the point group of symmetry of the crystal at the growth temperature. Therefore, the pulling operation leaves only the elements of symmetry of the crystal oriented parallel to the pulling direction. These symmetry elements are clearly reflected in the configuration and the form of the ridges. The ridges normal to the mirror plane and parallel to the pulling direction are symmetric in form. However, in the general form, the grooves on the side surface of the crystal may be non-symmetric.

The number and orientation of the ridges on the crystal surface can be determined by stereographic projection. This method is illustrated in Fig.34 where the upper row represent the projections of elements of symmetry of the high-temperature phase of the LN $\bar{3}2/m$ (D_3d) and six equivalent planes normal to the family $\{10\bar{2}\}$. The polar directions on the stereographic projection are parallel to the withdrawal axis. The second row contained only the elements parallel to the withdrawal axis and they represent axial groups for the selected withdrawal axis. To determine the number of ridges formed on the crystal, it is important

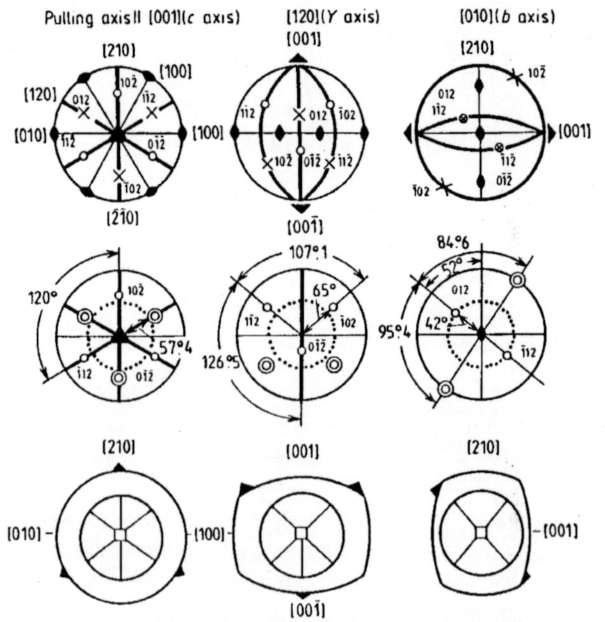

Fig.34 Graphic determination of growth ridge patterns for different directions of crystal growth. Explanation in the text (Niizeki et al, 1967).

to examine the angular relationships between the growth vector $\overline{G}$ and the normal plane. If the steady growth condition is reached and $\overline{G}$ is perpendicular to the melt surface, these angles are obtuse for the planes normal in the southern part of the projections sphere and corresponding grooves will form on the cylindrical part of the crystal. If vector G forms a small angle with the surface of the melt, all normal planes found in the southern direction from the line of the width in the northern hemisphere should form an obtuse angle with vector G. These lines are shown by dotted circles in the second row of the graphs. The small double circles like the small single circles represent the normal planes satisfying the conditions described previously. These normal planes correspond to ridges formed on the expanded part of the crystal. The mutual orientation of these ridges can be determined if the axial relationships are known and the given schemes are used. The lower row shows cross-sections of the crystals expected on the basis of the previous considerations, and also the position of the ridges on the expanded part of the crystal examined from the top which are shown on the internal side of the circles within the limits of the cross-section.

Information on the actual crystal growth conditions can be obtained by examining the morphology of the end surface of the crystal. At the conditions close to ideal, i.e. when the radial and axial temperature gradients and the rotation rate and withdrawal are selected optimum, the end surface is usually smooth, flat and slightly convex. As already mentioned, it is desirable to avoid rotation of the solidification front during crystal pulling. However, in practice, the surface is often coated with a large number of triagonal pyramids, discussed previously. Typical growth surfaces of the LN crystals, grown in the directions $[000\overline{1}]$,$[10\overline{1}0]$, and $[10\overline{1}\overline{2}]$, are shown in Fig.35 (Garmash, Znikina, Lazareva, 1972). This form of the surface results in the block structure of the crystal and a polycrystalline melt may form between the blocks. This structure of the crystal is characterised by high stresses. Defects make the crystal unsuitable for application in practice.

The front of growth of the crystals grown in the [0001] direction shows fragments of triagonal pyramids (Fig.35a). The grooves on the side surface and the growth front end in front sections. At the growth front, the fragments formed as a result of development and intergrowth of pyramids of growth of the $\{10\overline{1}\overline{2}\}$ faces, as already mentioned previously. The tangential development of elementary rhombic fragments of the growth pyramid $\{10\overline{1}\overline{2}\}$ results in the rib form in the direction $[10\overline{1}0]$ with the tip in $[000\overline{1}]$. At high growth rates determined by supercooling by several degrees, the form of the crystal becomes similar to that of the three-phase prism with truncated edges, and steps growth of the $[10\overline{1}\overline{2}]$ faces develops at the growth front (Fig.36).

Microscopic examination, carried out by Garmysh et al, shows that the $\{10\overline{1}\overline{2}\}$ faces grow in layers whose centre sometimes contains growth spirals. Summing up the above considerations, it should be noted that the morphology of the side surface of the crystals grown by the Czochralski method depends on the growth direction, and the morphology of the growth front depends on its curvature.

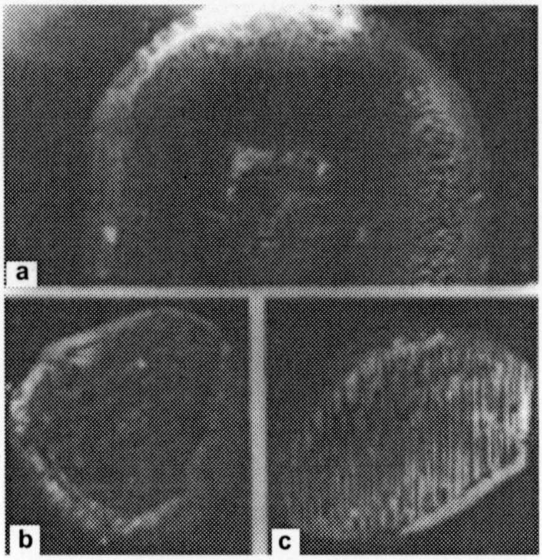

Fig.35 Growth front of lithium niobate crystals with high convexity of the surface grown in different directions: a) $[000\bar{1}]$; b)$[10\bar{1}0]$; c) $[10\bar{1}0]$, magnification ×5 (Garmash et al, 1972).

Special attention should be given to the formation of microscopic cavities in the form of circular pores or negative crystals in the form of dendrites in the central parts of crystals grown by the Czochralski method. The nucleation of cavities is usually associated with the variation of the growth conditions: expansion of the crystal is accompanied by the development of negative crystals, and pores appear as a result of rapid narrowing of the crystal. Impurities usually induce the formation of negative crystals without any change of the growth

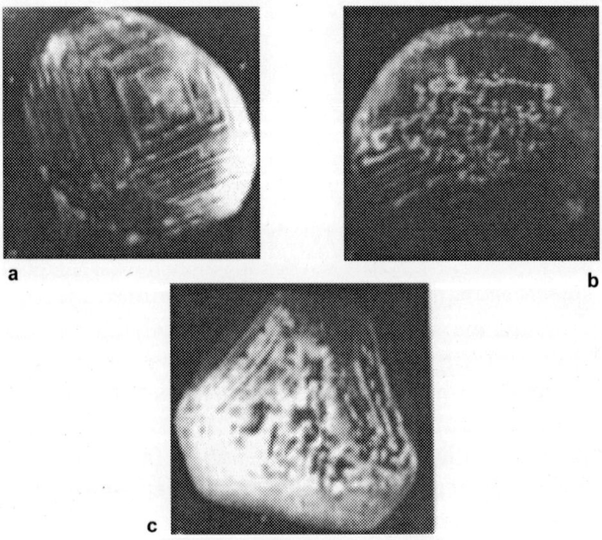

Fig.36 Surface of the growth front of lithium niobate crystals at different pulling rates and directions: a) low rate in the directions $[10\bar{1}\bar{2}]$ and b) $[000\bar{1}]$; c) high rate in $[000\bar{1}]$ (Garmash et al, 1972).

regime. The mechanism of formation of the negative crystals may be described as follows. During development of growth pyramids the impurity displaced into the melt builds up in the central most supercooled zone of the melt below the crystal. The temperature of the melt enriched with impurities in this local zone may be lower than the melting point of the pure crystal. An increase of the melt temperature may lead to melting out of the solidified low-temperature phase of the melt with the formation of circular pores.

12. Micromorphology of Crystal Surfaces

Attention will now be given to the micromorphology of crystals which is usually developed by etching the surface with a mixture of hydrofluoric and hydrochloric acids. In the lithium niobate the micromorphology makes it possible to develop a domain structure which can then be used to evaluate the degree of polarization of the crystal. Only monodomain crystals are used in practice.

Spontaneous polarization in lithium niobate is parallel to directions $+C$ and $-C$. Antiparallel domains on crystal cuts are developed by etching. The negative side of the (001) plane is etched at a higher rate and shows trigonal etch figures. The positive side requires a longer time for etching. Etch figures have the form of hillocks and etch pits. Niizeki et al, (1967) found that the opposite sides of the $(0\bar{1}0)$ plane, i.e. opposite ends of the Y-axis, can also be clearly separated by means of etching.

The relationship between the axial orientations of the antiparallel domains, developed by etching, is shown graphically in Fig.37. The graph shows that these two domains are linked by the axis of the second order, parallel to the [100] axis, i.e. they are normal to the mirror plane. Thus, the results show that the element of symmetry linking the domains polarised in opposite directions is the common element for point groups of ferroelectric and paraelectric phases.

In fact, the phase transition from the paraelectric to ferroelectric phase $\bar{3}2/m \rightarrow 3m$ observed by Niizeki (1967) has an axis of the second order. Determination of the positive direction of the Y-axis is important in producing a crystalline element for parametric generation of light.

A typical etch pattern on the C-plane of a polydomain crystal is shown in Fig.38. The density of hillocks varies over a wide range and its mean value is 10^4–10^5 cm^{-2} for crystals grown by the Czochralski method. Hillocks in polydomain crystals are distributed around the domain boundaries (see Fig.38). Due of their origin, the hillocks are linked with a source of limited length because in long-term etching some of the hillocks appear and disappear.

Detailed electron microscopic examination showed that the hillocks had the form a complex polyhedron and some of them have a flat tip. The linear dimensions of the tip vary from 0.1 to 1 μm, the base of the polyhedron has a size of 0.5–15 μm. Etching experiments showed that the flat tips of the hillocks relate to the positive surface of the domain and etch at the same rate. During etching the diameter of the central part of the hillock slightly increases by 0.3 μm. This is caused by more stable capacity to resist etching. During further etching a new hillock appears in the area of the previous hillock. On the basis

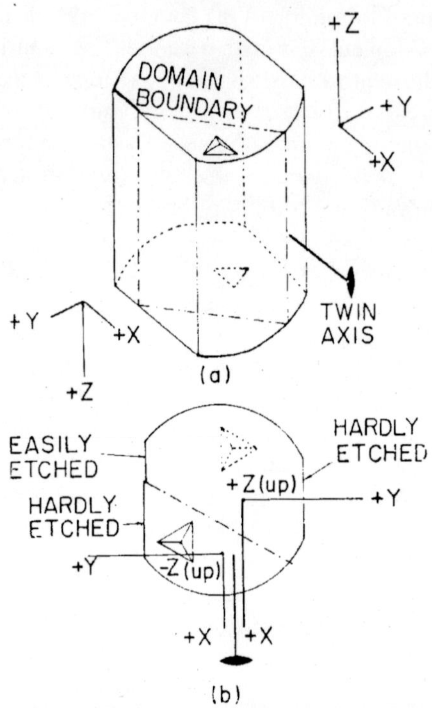

Fig.37 Relationship between antiparallel domains in lithium niobate crystals; a) perspective; b) projection on (001) plane (Niizeki et al, 1967).

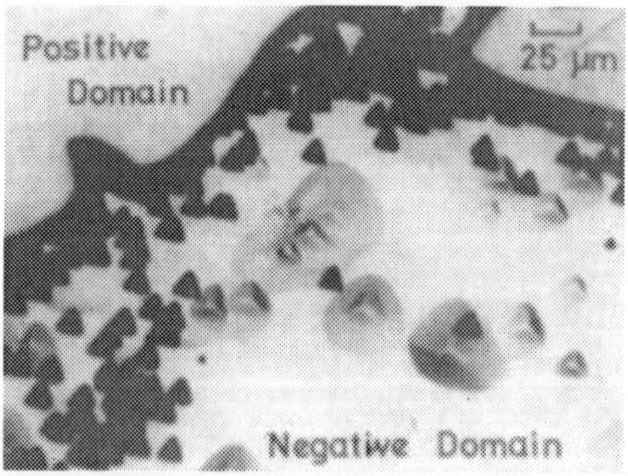

Fig.38 Typical etch patterns on the negative C-surface of a multidomain lithium niobate crystal. Dark triangles on the surface of the negative domain represent etch hillocks (Ohnishi, Iizuka, 1975).

of these experiments it may be concluded that the hillock has a needle-shaped core (centre) which is situated in the centre of the hillock and is parallel to the *C*-axis. The schematic explanation of the formation of a hillock is shown in Fig.39.

Long-time etching for 10–30 min makes it possible to explain the forma-

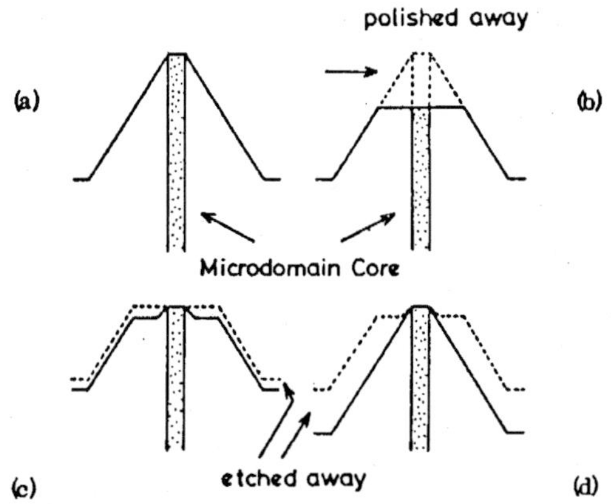

Fig.39 Schematic representation of cross-sections of cuts of hillocks, a) prior to etching; b) after mechanical polishing; c) after light etching for 10 s; d) after long-term etching (Ohnishi, Iizuka, 1975).

tion of etch pits on the positive surface *C*. The results of etching experiments showed two types of etch pits whose density was of the order of 10^5 cm^{-2}. Simultaneous etching of the positive and negative surfaces showed a close correspondence between the hillocks and etch pits on different surfaces of the same sheet. The experiments have also shown that the corresponding pit and the hillock originate genetically from the same source extended along the *C*-axis whose maximum length is approximately 300 μm. The second type of etch pits on the positive surface forms large areas which correspond approximately to large areas of the negative surface. This correspondence is observed even at a distance of approximately 2 mm. It is assumed that this type of etch pits is associated with the exit of dislocations. The exact correspondence of the etch pits on both sides of the sheet is difficult to determine because of the high density of the etch pits.

Of special interest are the etch patterns on the *y*-surface, one of which has the typical form shown in Fig.40. The complex pattern of the domains consists of coarse and smooth surfaces. The smooth surface corresponds to the positive *y*-direction, the coarse one to the negative direction. A typical feature of the domain structure of lithium niobate is the very complicated configuration of the domains. The needle-shaped domains are elongated along the *C*-axis in smooth and rough surfaces of the *y*-sheet. The positive needle-shaped domains, whose height increases in the rough area, are immersed in the matrix of the negative domains, and the negative needle-shaped domains, which form depressions on the smooth surface, are surrounded by the matrix of the positive domains. The density of the needle-shaped domains on the *y*-surface is 10^4 cm^{-2}. The positive needles are 200–400 μm long and less than 5 μm wide. The needle-shaped domains are concentrated at the domain boundaries which are almost perpendicular to the *C*-axis of the *y*-surface. The needle-like formations were also found by the methods of x-ray topography and polarization microscopy.

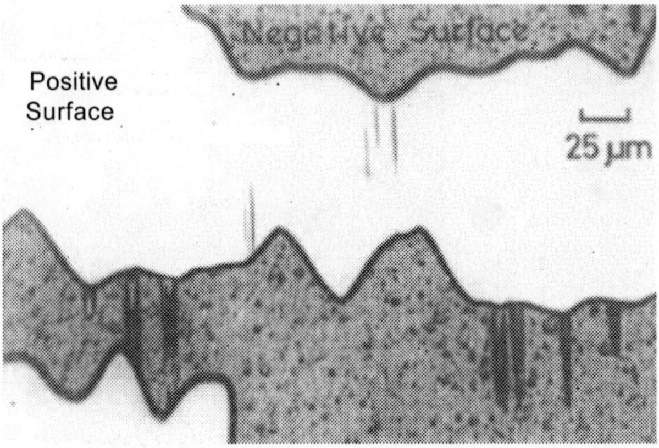

Fig.40 Typical etch pattern of the Y-surface of a multidomain crystal. Smooth and coarse regions correspond to the positive and negative Y-surface, respectively. Needle-shaped domains are parallel to the axis C and their traces are visible on both surface (Ohnishi, Iizuka, 1975).

Formations similar to needle-domains were found by x-ray diffraction in crystals of barium–strontium niobate ($Ba_xSr_{1-x}Nb_2O_6$) (Revkevich et al, 1979). These authors concluded that the monodomain crystals are not completely polarised in one direction and contain microdomains which are polarised in the reverse direction in relation to the crystal surrounding them.

The mechanism of formation of the microdomain has not as yet been explained. Revkevich et al (1979) also assumed that the local stresses, formed by scratching the surface of the crystal, changed the polarity of the microscopic region to the reversed polarity which is manifested in etching as a result of formation of a pit or a hillock, respectively. Figure 41 shows schematically the cross-section of a microdomain which is a source of both a hillock and an etch pit.

13. Electronic Exoemission of Lithium Niobate Pyroelectric

Thermoelectronic emission and thermally stimulated exoelectronic emission are the well-known phenomena of electronic emission which occur with the increasing temperature of the solid. Rosenblum et al (1974) found another type of thermally stimulated emission which occurs in pyroelectric materials. Electric fields, formed as a result of the thermally stimulated change of spontaneous polarization P_s of this material, caused exoelectronic emission from its surface. This phenomenon may be utilised for visualising the domain structure of a crystalline cut and for evaluating the degree to which the crystals is of the monodomain type.

It is well known that air breakdown takes place during a variation of the temperature of the polarised lithium niobate single crystal. The observed discharge phenomenon indicates that the thermally generated electric fields in certain pyroelectric crystals may be very strong and their strength is sufficient to cause field emission of the electrons.

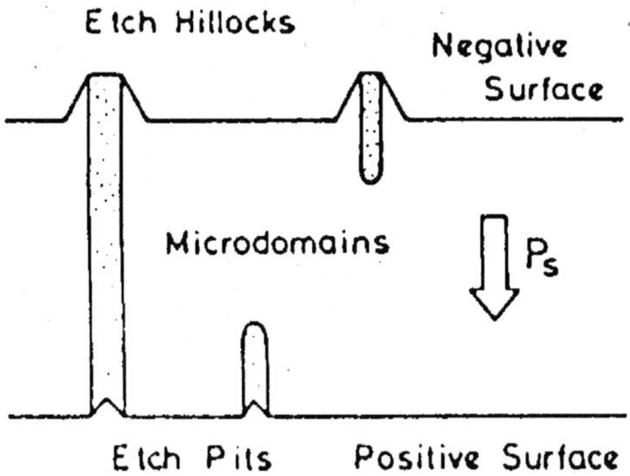

Fig.41 Schematic image of the cross-section of microdomains, etch hillocks and etch pits (Ohnishi, Iizuka, 1975).

This phenomenon was studied in vacuum on lithium niobate crystals. Fluorescent screens were used to visualise electronic emission from the surface. Spatial resolution power reached 300 µm. The crystals were assembled in a vacuum system using a copper holder of specimens with which the crystal was heated and cooled. The distance between the crystal and the detector surface was approximately 1–2 mm. The intensity of the emission current from the crystal surface depended on the relative dimensions C^- and C^+ of the domains extending to the surface. The typical image of emission from a polydomain surface, obtained during cooling, is shown in Fig.42a. The photograph of the same surface after etching in hydrofluoric acid is shown in Fig.42b. Comparison of these photographs shows that the pattern of thermal electronic emission corresponds to the domain structure developed by etching.

14. Formation of the Polar State of Crystals

Polarization of the crystals, i.e. transfer of the crystal to the monodomain state, is an important part of crystal technology. The crystals in the polydomain state lose their electro-optic and non-linear properties, and undesirable optical heterogeneities form in the crystals.

The polydomain state of the crystal was discussed in connection with the micromorphology of crystal surfaces. The domain structure of the crystal can be developed by surface etching methods. The domain structure and polarization mechanisms of lithium niobate were described in detail in a book by Kuz'minov (Kuz'minov, 1987). Here we present only general considerations essential for producing the monodomain state and discuss in greater details the methods of determining the extent of polarization.

Polarization is carried out at temperatures close to the Curie temperature. For crystals close to congruent composition, the Curie temperature T_C is approximately 1170°C. High temperature and also the presence of an alkaline metal greatly complicate the polarization process. The group polarization method is used in industry. A platinum sheet is placed on the floor of a muffle furnace

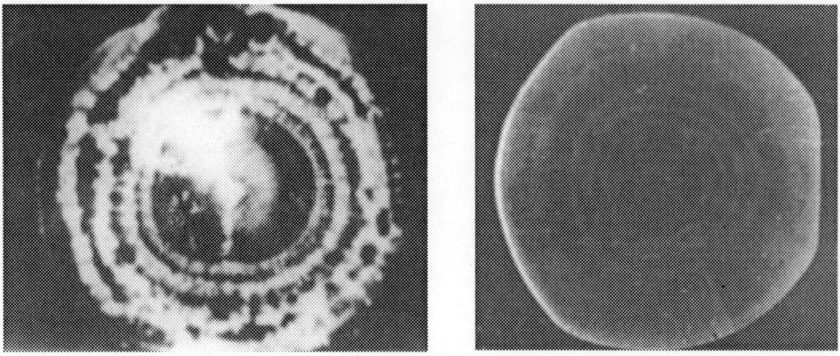

a b

Fig.42 Emission of a polydomain lithium niobate crystal: a) photograph of emission during cooling of the crystal; b) photograph of the crystal after etching in HF acid. Crystal diameter 15mm (Rosenblum et al, 1974).

and is covered with a layer of the lithium niobate charge. Crystals are placed in the vertical position on the cover and platinum wires, playing the role of positive electrodes, are connected to crystal seeds. At 1170°C a voltage is applied to the crystal. The voltage is determined by the diameter and length of the boule and also by electrical conductivity. Therefore, each crystal has its own power source which maintains constant the current flowing through the crystal. After holding at constant temperature and current for 1 hour, the temperature gradually decreases to 950–1000°C and the current is then switched off. When the current is switched off at lower temperature (700–800°C), the crystal becomes brown as a result of the reduction of niobium and a loss of oxygen, and its electrical conductivity also increases. This phenomenon plays a positive role in the production of devices based on surface-active substances: pyrocharges, formed at small fluctuations of temperature, are rapidly neutralised. It is well known that the field, formed as a result of the pyroeffect, causes a breakdown of the crystal on the surface and the device is damaged. It should also be mentioned that securing the positive electron in the form of wire to a seed crystal may lead to the heterogeneity of the field in the volume of the crystal or even to its breakdown in the central part, and, consequently, the crystal will not be completely of the monodomain type. To prevent undesirable effects, it is possible to remove the upper part of the crystal and place on the flat surface a platinum electrode on the charge layer to prevent spallation.

In carrying out polarization at high temperature, the crystal is also annealed and this greatly improves its optical homogeneity. Monodomainization of LN crystals is usually carried out at 1160–1210°C (Kuz'minov, 1987). Under these conditions, an increase of current density to 50 mA/cm^2 or an increase of the effect time leads to the appearance of a brown tint at the negative electrode which is then spread over the entire crystal. Colouring of the crystal is associated with the reduction of niobium and appearance of charged defects in the

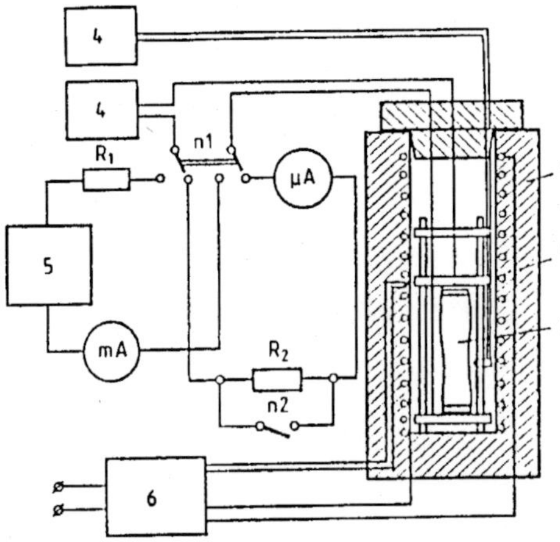

Fig.43 Scheme of an installation for polarization and high-temperature electrodiffusion annealing of LN single crystals: 1) low-temperature gradient furnace; 2) alundum crystal holder; 3) LN crystal; 4) devices for recording current and temperature; 5) constant voltage source; 6) programme temperature regulator; R_1 – operation switcher; R_2 – measuring resistance shunt (D'yakov, 1982).

crystal. The circuit diagram of equipment for high-temperature monodomainization is shown in Fig.43.

Polarization of the crystal can be carried out during its growth in growth equipment, but this method is not used widely because of problems in controlling the process and the effect of the current passing through the crystal on crystal growth. In growing crystals in an induction furnace initial difficulties may arise due to the passage of strong currents through a closed system causing the crystal to melt.

We believe that the most promising method of producing monodomain crystals is the one which utilises a relatively large thermal zone above the surface of the melt with a low temperature gradient. Formation of the single domain state is supported by ensuring the maximum possible distance of the Curie isotherm from the phase boundary and by its flat or slightly convex form. Long holding of the crystal at temperatures above the Curie point improves its homogeneity, and a flat Curie isotherm facilitates the monodomainization process because the volume of the crystal in which the structure is of the single domain type increases with increasing ratio of the axial and radial gradients. It should be mentioned that these two requirements contain an internal contradiction from the viewpoint of experimental execution (Ivanova, 1979).

For production technology of lithium niobate crystals it is very important to have a simple and reliable method of determining the monodomain state of the crystal. It is important to know whether the crystal is completely of the monodomain type of not. Only monodomain crystals are used in instrument making. One such method is the piezoelectric resonate method proposed by the author of this work (Kuz'minov, 1975).

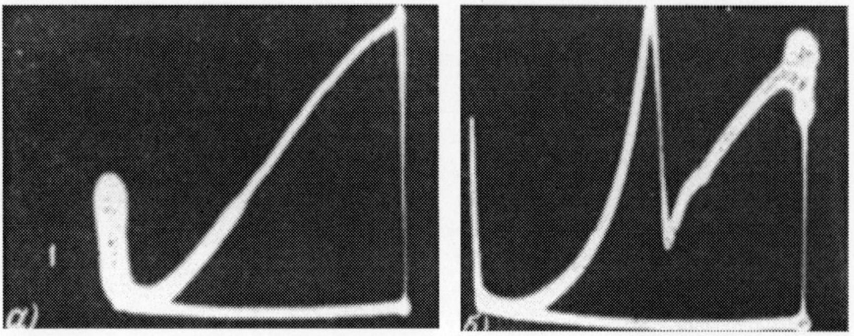

Fig.44 Amplitude–frequency characteristics of a lithium niobate crystal: a) multidomain; b) monodomain (fully polarized) (Kuz'minov, 1975).

The method can be used for qualitative determination of the extent to which the crystal is of the monodomain type. For measurements, it is not necessary to carry out optical treatment of the crystal and it is sufficient to have parallel ends approximately perpendicular to the direction of the spontaneous polarization axis. Crystal boules 2–3 inches in diameter and 100–120 mm long can be examined using a special holder.

The monodomain state of the crystal is determined from the peak of the piezoelectric resonance. This method is based on exciting piezoelectric vibrations in a monodomain crystal; in a polydomain crystal, vibrations cannot be excited. The amplitude–frequency characteristics (AFC) of the examined crystal is observed on the screen of a device for measuring the frequency characteristics. Figure 44 shows the AFC of polydomain and monodomain crystals of lithium niobate. The height of the AFC peak depends on the measurement conditions (plane parallelity of the end surfaces, the angle of inclination of the axis of spontaneous polarization, the tightness of contact of the electrodes with the crystal). The AFC can also be recorded during monodomainization. The monodomain state of the crystal can be then examined in greater detail on the basis of the electro-optical effect (magnitude of the halfwave voltage).

15. Determination of the Direction of the Crystallographic Axis

There are two universal methods of determining the crystallographic orientation of a boule: the Laue method of back photography, and determination of Bragg angles in a diffractometer. The presence of ridges on the side surface of a boule makes it possible to determine the crystal orientation. As mentioned previously, the crystal with zero orientation, grown along the optical axis *C* (*Z*), contains three ridges on the side surfaces situated at an angle of 120° in relation to each other. They appear to form traces of intersection of symmetry planes with the side surface of the crystal. Figure 45 shows elements of the symmetry of the point group 3*m* which includes the lithium niobate. The graph shows that the crystallographic axis X_2 is positioned in one of the planes, the axis X_1 is perpendicular to the X_2 and X_3 (*Z*).

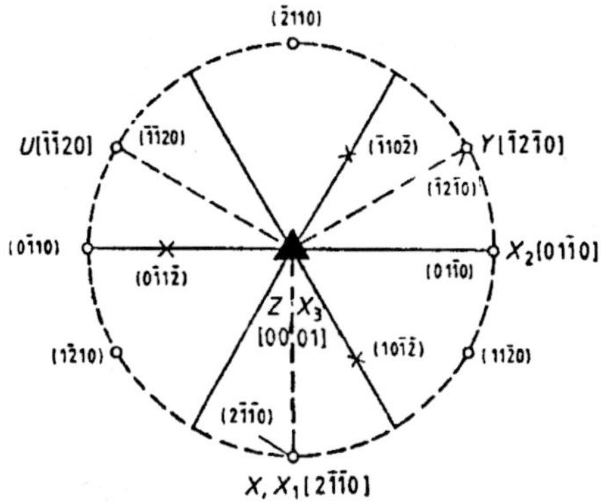

Fig.45 Elements of the symmetry of the point groups 3m to which the lithium niobate crystal belongs.

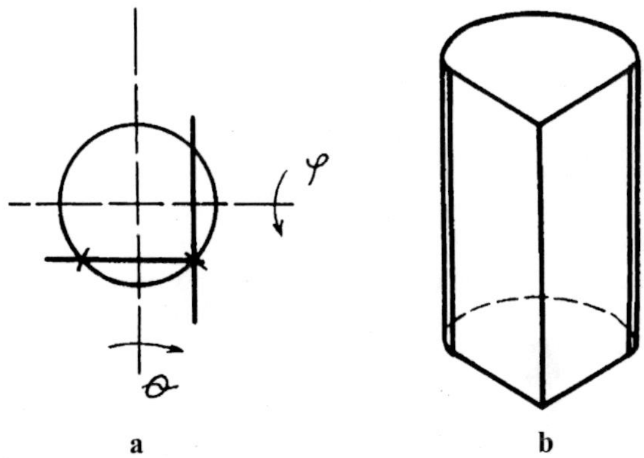

Fig.46 Diagram of the orientation with respect to striations on the side surface of the crystal.

The crystal is polished from the ends and is faceted, taking into account the position of the grooves of the crystal, shown in Fig.46. A polarization microscope and a Fedorov table (Melancholin 1970) are used for optical orientation. To orient the crystals up to 30 mm long, it is recommended to use an insert with a glass bottom on which crosshairs is made. The crystal is in the position in which the planes of its faces are parallel to the crosshairs. The crystal is rotated around two mutually perpendicular axes until the conoscopic pattern is transferred to the central part of the microscope eyepiece. In this case, the optical axis of the crystal coincides with the axis of the microscope. Rotation angles θ and ϕ, shown in Fig.46, are determined along the dial of the table taking into account the rotation direction. Actual rotation angles are determined by dividing the measured angles by the refraction indices of the crystal, approximately by 2.2. It should be mentioned that for crystals longer than 30 mm, the angle measurement range is restricted to 5–6°, and at high angles the beam leaves the crystal. To ensure that the optical axis *C* of the crystal is perpendicular to the end of the boule, the crystal is rotated to the opposite side during grinding. The accuracy of determining the angles is 30′. The exits of the axis X_1 and X_2 make it possible to determine the grooves present on the side surface of the boule; the error of such a determination may be several degrees.

Of special interest is the orientation of a lithium niobate crystal for parametric light generation. In this case, it is important to take into account the direction of the *Y* axis. The direction of the polarizing field is selected in such a manner that the boule axis is positioned in the negative quadrant *YZ*. This is important for application in the non-linear optics in order to maximise the non-linear optical coefficient. The effective coefficient is expressed as follows (Byer et al, 1974):

$$d_{\text{eff}} = d_{31}\sin\theta + d_{22}\cos\theta\sin 3\phi, \qquad (37)$$

where θ and ϕ are the angles formed by the axis of the boule and the axes *Z* and *X*, respectively, and the signs of the non-linear coefficients d_{31} and d_{22} differ. The crystallographic direction $[01\bar{4}]$, situated in the fourth quadrant, is closest to this direction. The direction $[01\bar{4}]$ is positioned in the plane of symmetry together with the axis X_2, [010] and forms with the axis *Z*, [001], an angle of 128°, and with the axis X_1, $[\bar{2}10]$ an angle of 90°. The correct selection of the direction $[01\bar{4}]$ is confirmed by the positive direction of the axis *Y* [110], which is determined from the positive piezoeffect. Figure 47 shows the stereographic projection of a lithium niobate crystal indicating the previously mentioned crystallographic direction, and Fig.48 shows the orientation of the crystal in relation to the crystallographic axis.

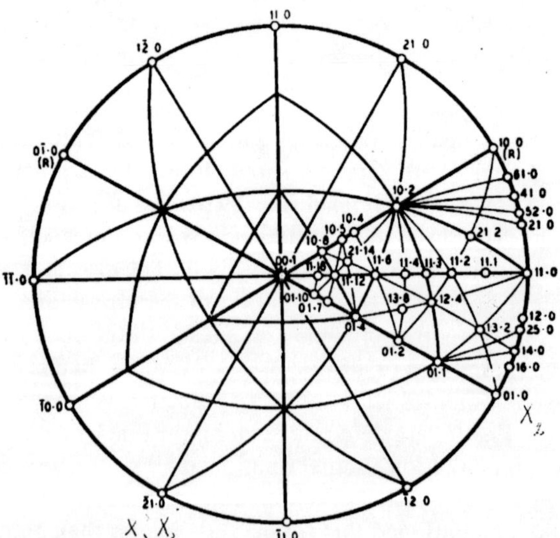

Fig.47 Stereographic projection of a lithium niobate crystal along the polar axis.

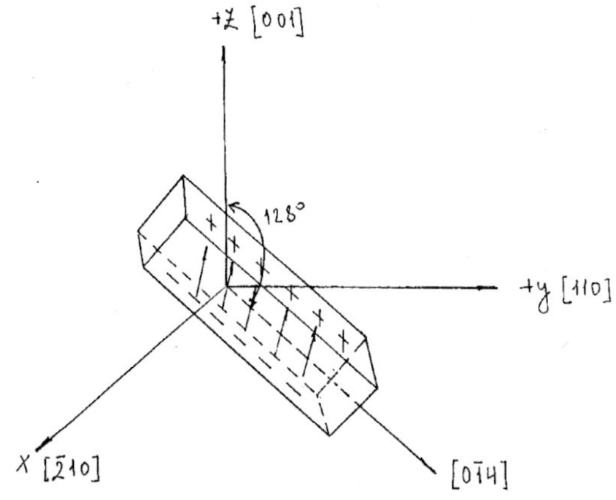

Fig.48 Schematic image of the orientation of a crystalline element of LN cut out along the $[01\bar{4}]$ direction, for parametric light generation.

16. Optical Methods of Examining Crystal Quality

To increase the yield of suitable components made of single crystals and make production economically efficient, it is necessary to determine the quality of the crystal in each processing stage. After growing the crystal and polarization the crystal is visually examined and the dimensions, form and morphology of the surface are recorded. This must be carried out in order to compare the results with the growth conditions and make appropriate corrections in the technological process.

More detailed information is provided by examining the crystal with a narrow beam of the light of an illuminator of a microscope or laser. Figure 49 shows a diagram of such a device which makes it possible to examine crystal defects both visually and record them on a film. On the basis of light scattering in the bulk of the crystal, this device makes it possible to determine the presence of domain boundaries, the block nature of the crystal, a second phase inclusion (for example, lithium triniobate) and also an inclusion of gas bubbles and the container metal (platinum). The presence of the Tindal cone characterises the scattering of light by finely dispersed particles of the second phase. The passing beam of the He–Ne laser in the ideal crystal becomes invisible and is reflected only on the side walls. The resultant information is of considerable interest and makes it possible to draw conclusions on subsequent technological stages: remelt the crystal, remove the defective part of the crystal, or carry out additional monodomainization .

16.1 Determination of birefringence in crystals of the basis of the conoscopic pattern

The simplest investigations of the optical heterogeneity of the crystal include

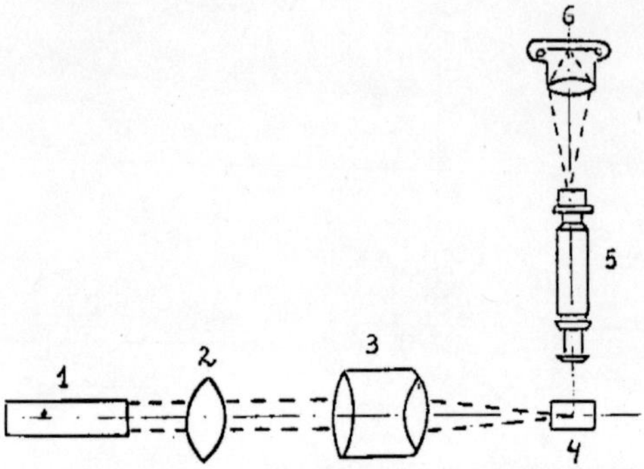

Fig.49 Diagram of equipment for examining the scattering of light in crystals: 1) He–Ne laser with $\lambda = 0.632$ μm; 2) condenser; 3) focusing microobjective; 4) examined crystal; 5) microscope barrell; 6) camera.

the determination of the birefringence of the crystal on the basis of its biaxial nature which can be determined quite easily from the conoscopic pattern in a polarization microscope in crossed polarisers, Fig.50.

The relationship between the biaxial nature of the crystal and its birefringence along the optical axis of the microscope is described by an approximate equation (Idenbom, Tomilovskii, 1958):

$$\sin\Omega = (\Delta / \Delta_0), \tag{38}$$

where 2Ω is the true angle between the optical axes of the crystal, $\Delta = n_1 - n_2$ and $\Delta_0 = n_e - n_o$ is the induced and natural birefringence, respectively.

In conoscopic examination it is necessary to measure the angle 2ε formed by the beams extended to the air and passed through the crystal in the direction of its optical axis. In transition to the true angle between the axes Ω it is necessary to introduce a correction for the refraction index on the examined crystal:

$$\sin\ \Omega\ =\ 1/n_0\ (\sin\ \varepsilon) \tag{39}$$

Conoscopic patterns are examined in a polarization microscope. The meas-

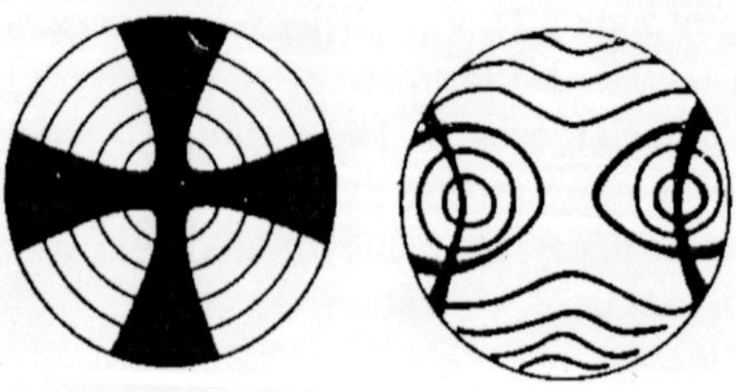

a b

Fig.50 Patterns observed in converging polarised light: a) uniaxial crystal; b) biaxial crystal.

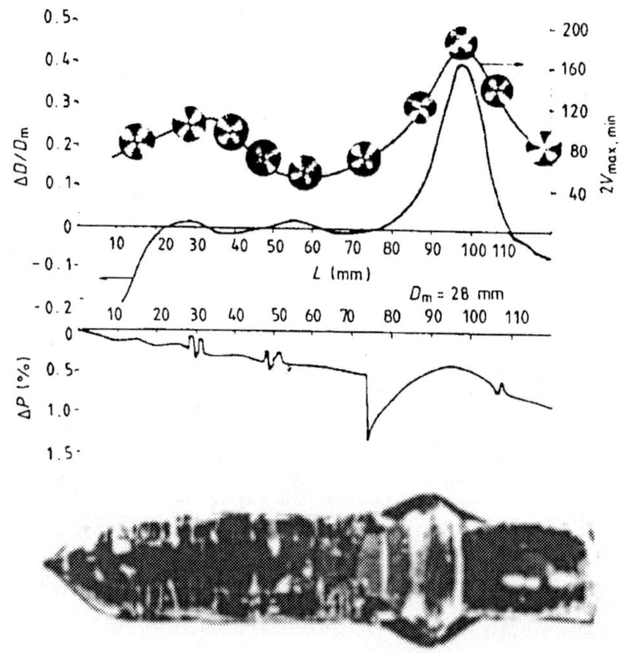

Fig.51 Example of the dependence of diameter of the crystal boule and the distribution of anomalous birefringence on the growth conditions; a) crystal boule; b) diagram of the distribution of anomalous birefringence, $(2\,V_{max})$ min; deviation of the crystal diameter $(\Delta D/D_m)$ and power of the hf generator (ΔP) in relation to the values in the stationary crystal growth regime (Buraches 1981).

ured distance d of exit of the optical axis is linked with the angle 2ε by a reference phlogopyte crystal sheet with $2\varepsilon = 10°30'$ and $n_o = 1.55$. The Mallar's constant, determined in this manner, is $R = 0.0355$ (Kuz'minov, 1975). Induced birefringence is determined from the previous equations:

$$\Delta = \Delta_0\ (R^2\ d^2/n_0^2) \qquad (40)$$

From the induced birefringence, we can determine the mechanical stresses in the crystals, as carried out by the author for the lithium niobate crystal. The stresses reach 50 kgf.

As an example, the relationship between birefringence in a lithium niobate crystal, the variation of its diameter and the heating power of the generator is shown in Fig.51.

16.2 Characterisation of crystals using an interferometer

Optical interferometers, in particular, the Mach–Zender interferometer, are used for more detailed examination of the optical perfection of crystals. However, the use of these devices imposes stringent requirements on the optical treat-

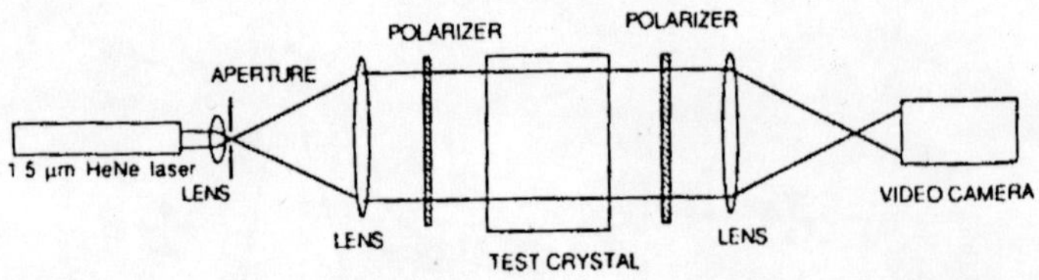

Fig.52 Diagram of a birefringence interferometer (Henningsen, Singh, 1989).

ment of the surface of the crystalline element. For example, its planeness should be in the range of a fraction of the wavelength of the light used for examination. To reduce the requirements on optical treatment of the crystals, Hennigsen and Singh (1989) used a birefringence interferometer whose diagram is shown in Fig.52. The examined crystal is placed in the interferometer in such a position that its optical axis forms an angle of 45° with the polarization plane of the beam. In this case, the incident beam is divided into the ordinary and extraordinary beams which interfere at the exit from the second polariser. The interference pattern may be transferred to a video monitor.

The phase difference ϕ between the two beams is determined by the equation

$$\phi = n_o L - n_e L = L(n_o - n_e). \quad (41)$$

The crystal positioned along the path of the beam causes distortions in the refractive index $\Delta(n_e - n_o)$. The equation (41) can be written in the form

$$\phi + \Delta\phi = L[n_o - n_e + \Delta(n_o - n_e)] \quad (42)$$

From equation (42), we obtain changes in the phase $\Delta\phi$ as a function of $(n_o - n_e)$

$$\Delta\phi = L\,\Delta(n_o - n_e) \quad (43)$$

During the variation of $\Delta\,(n_e - n_o)$ throughout the entire aperture of the device corresponding changes take place in $\Delta\phi$. The variation of $\Delta\phi$ by the value 2π produces a single interference fringe on the screen. From the number of fringes we determine the phase difference and, consequently, the variation of birefringence. The optical heterogeneity is determined from the equation

$$\text{grad}\, n = \Delta(n_o - n_e) / L. \quad (44)$$

The Mach–Zender interferometer is also to sensitive to variations of crystal length ΔL. These variations changes occur when both surfaces through which the laser beam passes have an imperfect planeness or are non-parallel. This special feature of the interferometer complicates interpretation of the interference pattern.

In the birefringence interferometer, the variation of the phase $\Delta\phi$ as a result of the variation of crystal length ΔL is determined by the equation:

$$\Delta\phi = (n_o - n_e)\Delta L = \Delta n \, \Delta L, \qquad (45)$$

i.e. by the quantity of the second order of smallness. For the Mach–Zender interferometer, the variation of the phase as a result of the variation of crystal length is determined from the equation

$$\Delta\phi = L\,\Delta n. \qquad (46)$$

Thus, the birefringence interferometer is less sensitive to the non-planeness of the crystalline element and reflects the optical heterogeneity in the bulk of the crystal.

In uniaxial crystals of the lithium niobate type, the extraordinary refractive index depends on the direction of passage of light and it can therefore be assumed that the block structure of the crystal (subgrain) will be visible as optical distortions in the crystal. The ordinary refractive index does not depend on the direction of propagation and remains constant. If it is assumed that the interference fringes are caused by the angular misorientation of the blocks in the crystal bulk, we can calculate the mean angular deviation of the lattice along the optical beam. The angular misorientation of the lattice $d\theta$ leads to the appearance of Δn which can be computed using modified equations proposed by Yariv (Yariv, 1976):

$$d\theta / dn_e(\theta) = 2n_e^{-3}(\theta) / (n_e^{-2} - n_o^{-2}) \sin 2\theta. \qquad (47)$$

The integral lattice misorientation along the path of the beam by 0.1° corresponds to three interference fringes on the monitor screen.

16.3 Optical examination of crystals plates for surface-acoustic waves

Optical examination of lithium niobate plates for surface-acoustic waves (SAW) has its special features. The diameter of the plates varies up to 3 inches, thickness 0.5 mm, and they are oriented under an angle of 128° in relation to the *Y*-axis. One surface of the plate is polished, the other one is ground. To examine defects over the entire surface of the plate, it is necessary to expand the parallel beam to the plates dimensions. On the other hand, high birefringence of lithium niobate and the small plate thickness make it necessary to use strictly monochromatic light for resolving the interference fringes of a high order. The diagram of the interferometer is similar to that shown in Fig.53 (Born et al, 1982).

From the background surface of the plate the light is scattered almost without any change of its polarization. This scattered light is observed by means of the second polarization filter with the transverse polarization direction. The maximum image of the interference fringes is observed when the sheet it rotated

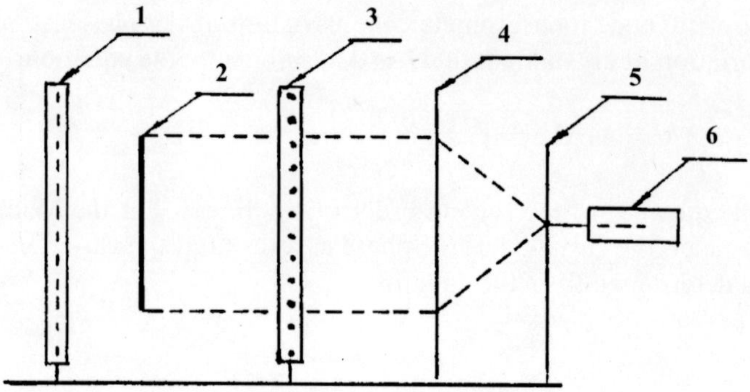

Fig.53 Schematic of an optical device for examining ferroelectric crystal plates using a passing orthoscopic laser beam (Born, Willibald, Veith, 1982). 1) polarizing filter || LN crystal; 2) crystal plates; 3) polarizing filter ⊥; 4) lens system ||; 5) lens system ⊥; 6) laser.

to the position in which the directions of polarization of the ordinary and extraordinary beams form an angle of 45° with the polarization direction of the filters P_1 and P_2 (Fig.53), respectively. If the ground side of the sheet is turned towards the light source, a conoscopic image is obtained. The cone is situated on the *a* axis and on the axis of the centre of the sheet. The conoscopic image shows the position of the optical axis of the sheet and this may be used as an additional verification of the accuracy of orientation of the sheet (see Fig.54a,b).

In the previous section, it has already been mentioned that the polarization methods of investigation make it possible to evaluate the misorientation of the blocks in the crystal bulk. The block boundaries determine the difference in the orientations of the optical axis of two parts (*a* and *b*) of the same crystal divided by this boundary. On the other hand, birefringence depends on the angle β between the incident beam and the optical axis of the crystal. Therefore, during passage through the crystal the ordinary and extraordinary beams do not interfere in the same phase in part *a* and part *b* of the crystal. In addition, these phases depends strongly on the length of the path of the beam through the crystal lattice equal to 0.5 mm.

The lithium niobate crystals are characterised by the uniaxial ellipsoid of

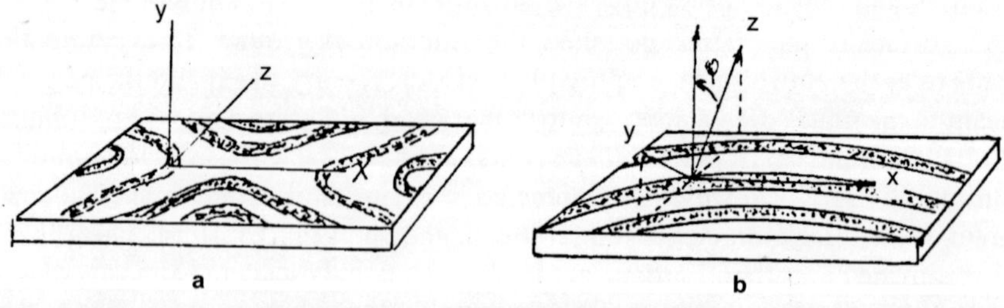

a b

Fig.54 Schematic of a conoscopic interference pattern in relation to the direction of crystallographic axis: a) Y is normal to the sheet plane; b) axis Y is directed at an angle of 128° to the sheet plane (Born et al, 1982).

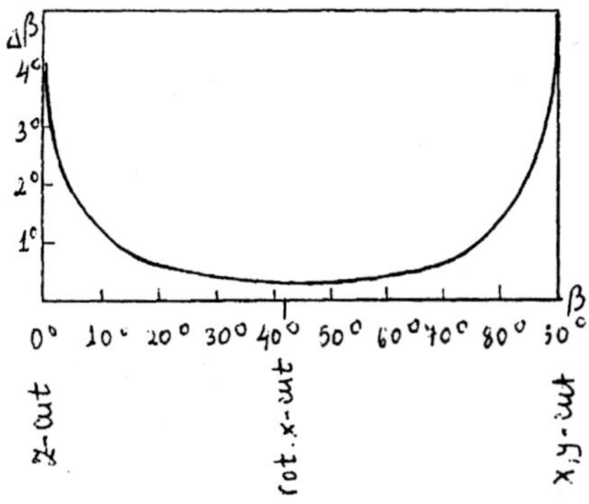

Fig.55 Dependence of the angle of misorientation $\Delta\beta$ of two optical axis of adjacent sections of the crystal resulting in the maximum contrast of low-angle boundaries between them, calculated as a function of angle β (Born et al, 1982).

refraction parameters. Any central section of this ellipsoid shows the double refraction of the crystal perpendicular to this section. Using the equation of the ellipse

$$n_{e'}^2 = n_e^2 / (1 - E^2 \cos^2 \beta), \tag{48}$$

where $E^2 = (n_o^2 - n_e^2) / n_o^2$, n_o and n_e are the refractive indices of the ordinary and extraordinary beams, $n_{e'}$ is the component of the refractive index of the extraordinary beam in the crystal. The angle $\Delta\beta = \beta_a - \beta_b$ is the difference of the phase shift in π between the ordinary and extraordinary beams which can be calculated. This phase shift makes the small-angle boundary visible with maximum contrast. The diagram (Fig.55) shows the dependence of the angle of misorientation $\Delta\beta$ of the optical axis of two adjacent parts *a* and *b* of the crystal in relation to angle β. The graph was calculated for the beam pass length in a crystal of 0.5 mm and for a wavelength of $\lambda = 0.625$ μm, which corresponds to the refractive index of $n_o = 2.2932$ and $n_e = 2.20235$. The graph shows the dependence of the contrast of the small-angle boundaries on the angle of rotation of the sheet in relation to the *Y*-axis.

The polarization methods can also be used to observe ferroelectric domains in sheets. The contrast of domain walls is explained by localization on domain walls of electric charges below the Curie temperature which induces the mechanical stresses and cause birefringence. The electrical conductivity mechanism compensates the electric charges over a period of several months at room temperature and makes the domains visible.

The polarization methods can be used to determine the non-uniformity of plate thickness. The interference order $n + x$ (n – integral, $0 < x < 1$) of the extraordinary and ordinary beams in the examined direction of the crystal depends on birefringence $(n_o - n_{e'})$ at $n_o > n_{e'} > n_e$, and also on the beam path length in the crystal, as mentioned previously. The difference in the paths corresponds to the difference of the thicknesses of the specimen and is therefore observed by means of interference fringes. The difference of the thicknesses between two fringes can be computed from the equation

$$\Delta l \cong \lambda' n_e n_o / (n_o - n_{e'}), \tag{49}$$

where $\lambda' = \lambda/n$ is the true wavelength in the crystal. Equation (49) shows that in X- and Y-cuts of the crystal the difference of one interference fringe corresponds to a difference in the thickness of 7 μm and in the step of 15 μm (128°–Y)-cut. These measurements do not depend on the absolute thickness of the sheet. The thinned part of the sheet will be visualised by heating the crystal during investigations to 30–40°C. For this purpose, it is sufficient to touch the sheet with a thumb. The fringes will move in the direction to the thin part of the sheet because of the fact that the birefringence of the lithium niobate increases with increasing temperature. In this section, we have shown extensive possibilities of polarization methods of examining the quality of crystals in the final stage of the technological process.

17. Optical Properties of Reduced Lithium Niobate

The optical properties of pure and reduced lithium niobate crystals are of special interest because of the extensive application of this material and also due to the fact that niobium may be partially reduced in producing this material. The pure stoichiometric LN is optically transparent but reduction is accompanied by the formation of a deficit of oxygen so that the optical properties of the crystals greatly change and an absorption fringe appears in the spectrum in the range 2.5eV. Weiging et al (1987) determined more accurately the position of the absorption fringe of reduced LN and made several assumptions regarding its nature.

The methods of reduction and examination of the crystals of this type are of interest for practice and we shall therefore discuss them in greater detail. For investigations, crystals with a congruent composition were grown by the Czochralski method, oriented by the x-ray method, and sheets $10\times6\times1$ mm^3 were then cut. The thickness along the C axis was 1 mm. In the first stage, all sheets were annealed in oxygen at 1000°C for 10 h. These sheets were then referred to as 'non-reduced' and used for further experiments. Specimens with a different degree of reduction were produced by heating in a vacuum of 10^{-3} torr at different temperatures in the range 600–1100°C. After reduction, the colour of the specimens changed from colourless for the non-reduced specimens to dark brown for the highly reduced specimens.

Optical transmission of the specimens in the wavelength range 0.2–0.9 μm was recorded with a two-beam spectrometer in passing a non-polarised light beam along the optical axis of the sheet. Measurements were taken in air, at room temperature. The absorption coefficient α was computed from the equation given below, using the Newton–Ralphson method:

$$T\{[n(n+2)(n+1)^2+(1+4n+2n^2)k^2+k^4]\times \\ \times\exp(4\pi kd/\lambda)+[n(n-2)(n-1)^2+(1-4n+2n^2)k^2+k^4] \\ \times\exp(-4\pi kd/\lambda\}-16(n^2+k^2)=0 \tag{50}$$

where n is the refraction factor, d is the thickness of the specimen, λ is the wavelength of incident light, and k is the extinction coefficient which is linked with α by relationship:

$$\alpha = 4\pi k/\lambda. \tag{51}$$

Electrical measurements of conductivity and thermoelectric power in relation to temperature were taken for all reduced specimens.

The transmission spectra of the non-reduced specimens and specimens reduced to various degrees are shown in Fig.56. Transmission for the non-reduced specimens reaches 70% at a wavelength of 0.9 μm, and for the reduced specimens it decreases in relation to increasing degree of reduction. On the basis of the graph it may be concluded that the sharp edge of the transmissions spectrum on the non-reduced specimen gradually becomes diffuse with increasing degree of reduction and is transformed into a 'tail' at strong reduction. Attention should also be given to the increase of the displacement of the spectrum to its low energies with increasing degree of reduction. However, this shift is very small and can be observed only on the expanded wavelength scale, as shown in the insert in Fig.56.

The absorption factors as a function of the energy of incident photons for the non-reduced specimen and specimens reduced to various degrees are shown in Fig.57. The graph shows that the absorption factor α monotonically increases with increasing photon energy and the magnitude of this increase greatly increases on the side of high energies because of interband electron transitions.

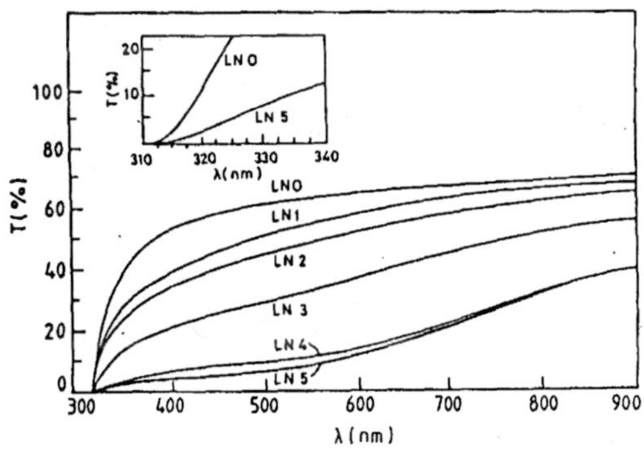

Fig.56 Transmission spectrum of a pure (LN0) LN crystal and of an LN crystal with various degrees of reduction (LN1–LN5). The insert shows the spectra of specimens (LN0 and LN5) on expanded wavelength scale (Dhar A., Mansingh A., 1990).

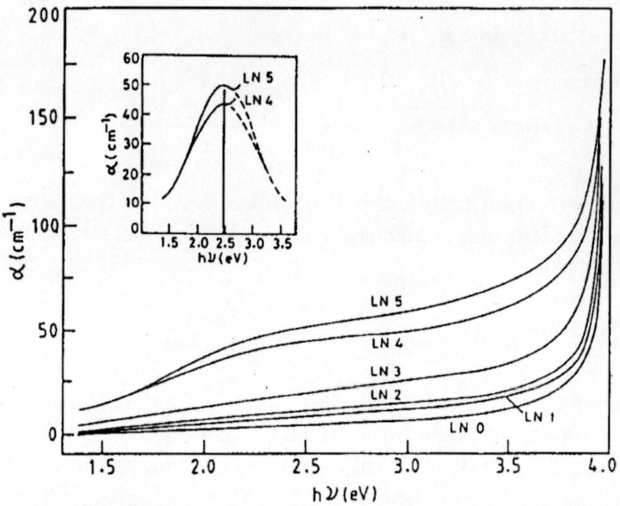

Fig.57 Absorption coefficient of LN crystals with various degrees of reduction as a function of photon energy. The insert indicates an absorption band of 2.48 eV of specimens LN4 and LN5 (Dhar A., Mansingh A., 1990).

For the strongly reduced specimens the centre of the width of the absorption band is close to 2.48 eV.

For the interband electron transitions in the vicinity of the fundamental absorption range, α is specified by the equation:

$$\alpha = B(h\nu - E_g)^r \tag{52}$$

where B is a constant, E_g is the interval between the optical bands, r is the index which can assume the values of 1/2 or 2 in relation to the type of transition for indirect and direct transition, respectively. Figure 58 shows indirect transitions in the reduced and non-reduced lithium niobate crystals. Intersections of the dotted straight lines with the x-axis gives the values of the interband transitions: for the non-reduced crystal $E_g = 3.78$ eV, which decreases to 3.45 eV for strongly reduced specimens. The insert in Fig.58 shows the variation of the width of the band for indirect transition in relation to reduction temperature.

Jin et al (1987) linked the absorption band in the vicinity of 2.48 eV with the presence in the reduced lithium niobate crystals of polarons of a small diameter which form in reducing niobium to Nb^{4+}. In accordance with the theory of Reik and Heese (Reik and Heese, 1967), the absorption factor α in this case is determined by the expression:

$$\alpha = (\sigma(0) / nc\varepsilon_0)\sinh(h\omega / 2kT) / (h\omega / 2kT)\exp(-h^2\omega^2 / 16WkT) \tag{53}$$

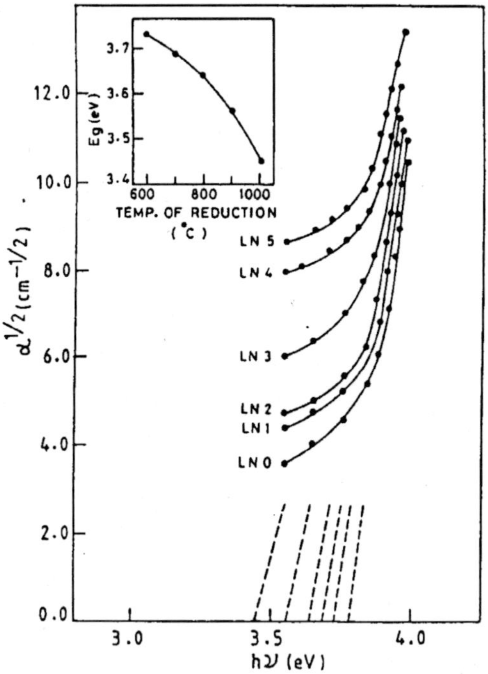

Fig.58 Dependence of $\alpha^{1/2}$ on photon energy $h\nu$ for indirect transitions of LN crystals with various degrees of reduction of LN. The inset shows the variations of the widths of the forbidden zone in relation to reduction temperature (Jin W., et al, 1987).

where $\sigma(0)$ is dc conductivity, ε_0 is the free-space permittivity, c is the velocity of light, W is the activation energy for dc conductivity, ω is the angular frequency of incident light, T is absolute temperature and h is Planck's constant. The optical absorption band is linked with the activation energy of conductivity by the equation:

$$E_{opt} = 4W \tag{54}$$

The total width at half height of the maximum of the absorption band $\Delta_{1/2}$ is given by the equation:

$$\Delta_{1/2} = 8(WkT) \tag{55}$$

The last two equations may be useful for characterising the reduced crystals. The activation energy of conductivity, computed from photon energy, corresponding to the maximum absorption band, is 0.62 eV, the measured value is $W = 0.51$ eV. The absorption factor at some frequency can be used to evaluate conductivity using equation (53) for the known values of the refraction index

Table 5 Electrical parameters of LN crystals (Jin, Zhon 1987)

Parameter	Measured	Theoretical
W (eV)	0.51	0.62
$\sigma(0)$ (ohm^{-1} cm^{-1})	1.58×10^{-10}	5.1×10^{-10}
$\Delta_{1/2}$	1.4	1.01

and activation energy. The experimental and calculated values of the reduced lithium niobate are presented in Table 5. It should be mentioned that the values of electrical conductivity and activation energy predicted on the basis of the Reik–Heese theory are quite close to the measured data. Consequently, the experimentally measured conductivity and activation energy can be used to estimate the absorption factor, using equation (53), assuming that the refraction factor does not depend on the frequency of incident light. The last assumption does not correspond to the actual situation and this results in a large difference between the measured and calculated values of the absorption factor.

18 Structural Defects in Lithium Niobate Crystals

18.1 Schottky and Frenkel defects

We shall examine possible variants of the formation of defects according to Frenkel since the formation of Schottky defects in the crystal structure of lithium niobate is disadvantageous from the energy viewpoint, as shown by Donnerberg et al, 1989. Frenkel defects form by the following reactions:

$$\text{"LiNbO}_3\text{"} \rightarrow V'_{Li} + Li \tag{56}$$

$$\text{"LiNbO}_3\text{"} \rightarrow V^{5'}_{Nb} + Nb^{5}_{i} \tag{57}$$

$$\text{"LiNbO}_3\text{"} \rightarrow V^{..}_{0} + Nb^{5}_{i} \tag{58}$$

The absolute values of the energy of formation of the defects are presented in Table 6. As a result of the high energy of formation of Nb, Frenkel defects do not form in the LN structure. On the other hand, Li-defects have a high probability of formation as a result of the relatively low formation energy and also due to the presence of vacancy sites in the structure of LN comparable with the dimensions of lithium cations. Li-defects are also responsible for ionic electrical conductivity in the LN crystals. The measured activation energy of 1.1 eV is close to the calculated value. Table 6 also shows that Frenkel defects in the structure of $BaTiO_3$ are less likely to form in comparison with lithium niobate.

Table 6 Energy (eV) of formation of Schottky and Frenkel defects in $LiNbO_3$ and $BaTiO_3$ crystals (Donnerverg et al, 1989).

$LiNbO_3$		$BaTiO_3$	
Schottky defects			
Li_2O	1.94	BaO	2.58
Nb_2O_5	2.85	TiO_2	2.90
$LiNbO_3$	3.91	$BaTiO_3$	2.29
Frenkel defects			
O^{2-}	3.42	O^{2-}	4.49
Nb^{5+}	6.26	Ti^{4+}	7.56
Li^+	0.93	Ba^{2+}	5.94

18.2 Defects formed when stoichiometry is disrupted

Defects inside a crystal can also form if the stoichiometry of the crystal composition is disrupted. This may be achieved varying the composition of the initial melt. The following relationships were determined by experiments:

1. Basically, only a shortage of Li_2O can occur.
2. The density of LN increases with increasing deficit of Li_2O.
3. The mechanisms of charge compensation of the deficit of Li_2O can be: V^{5-}_{Nb} and Nb^{4+}_{Li} (Peterson, Carnevale, 1972).

Initially, we examine the stoichiometric equilibrium on the side rich in Li_2O. The following three reactions in the solution can be proposed:

$$"LiNbO_3"+Li_2O \rightarrow 2LiNbO_3 + 2V_{Nb}^{5-} + 5V_O^{2+} \quad (59)$$

$$"LiNbO_3"+Li_2O \rightarrow LiNbO_3 + Li_{Nb}^{4-} + 2V_O^{2+} \quad (60)$$

$$"LiNbO_3"+Li_2O \rightarrow 2Li_i + O_i^{2-} \quad (61)$$

with defects formation energies calculated by Peterson and Carnevale, of 17.66, 5.82, and 4.71 eV respectively. Comparison of these values shows that reaction (61) is preferred from the energy viewpoint. However, an excess of lithium oxide may disturb the LN lattice and cause the formation of another phase, preferably Ni_3NbO_4. Thus, the implantation of a surplus of lithium oxide in lithium niobate crystals is unlikely to take place.

Therefore, we shall examine the shortage of lithium oxide assuming diffusion of Li_2O on the surface of the crystal:

$$"LiNbO_3" \rightarrow Li_2O + 2V_{Li}^{-} + V_O^{2+} \qquad (62)$$

$$"LiNbO_3" \rightarrow 3Li_2O + 4V_{Li}^{-} + 2Nb_{Li}^{4+} \qquad (63)$$

$$"LiNbO_3" \rightarrow 3Li_2O + 4V_{Nb}^{5-} + 5Nb_{Li}^{4+} \qquad (64)$$

where the last reaction can be modified:

$$"LiNbO_3" \rightarrow 3Li_2O + 4\left(V_{Nb}^{5-}\ Nb_{Li}^{4+}\right)^{-} + Nb_{Li}^{4+} \qquad (65)$$

The calculated energies of formation of these defects are as follows: 5.82, 4.56, 15.17 and 10.67 eV. The authors assume that the formation of oxygen vacancies by the model (62) is ineffective from the energy viewpoint. Therefore, the deficit of Li_2O can be compensated only by the formation of defects in the cations sublattice. The increase of the density of the crystal with increasing shortage of Li_2O is also consistent with the reactions (63)–(65). Thus, the oxygen sublattice is not disrupted and the surplus of niobium is sufficiently dissolved inside the remaining LN by substitution of lithium sites. Therefore, regardless of the formation of cation vacancies, the mass per unit volume increases, i.e. the crystal density increases. As an example, we examine the reaction described by equation (63). For each molecule of Li_2O, which diffused outside the crystal, the crystal density increases by 1.3 [(92.9 – 6.90 – (4 × 6.9)] = 19 atomic mass units. Although the models (64) and (65) are in agreement with (57) and (58), there are unlikely to be valid from the energy viewpoint. The difference in the energies of the reactions (65) and (63) is around 6 eV per Li_2O molecule. This result is not unexpected and was qualitatively predicted by Smit (1984). The contradiction with the calculated energies and the values observed for Nb and Li sites in diffraction investigations can be resolved if it is assumed that the LN contains more regions with the ilmenite-like structural sequence. The crystal structures of the ilmenite and LN are very similar and differ only by their position of the cations, as shown in Fig.59. The lattice energy, calculated per unit cell of the ilmenite- type, is only 0.1 eV higher than the corresponding energy of perfect LN. Thus, both structure can mix in the LN. The existence of the ilmenite-like LN has been confirmed by experiments carried out by Kumada et al (1985) who produced LN with an ilmenite structure using $NaNbO_3$ as the starting material.

18.3 Defects formed in reduction of lithium niobate

Experiments show that the reduction of LN crystals is accompanied by an increase in the crystal density and diffusion on the outside of O_2 and also possibly Li_2O, although the ratio $[Li_2O] / [O_2]$ remains unknown and depends on the

$LiNbO_3$	Ilmenite
B	B
A	A
B	A
A	B
B	B
A	A
B	A
A	B

Fig.59 Distribution of cations in the crystal lattice of LN and ilmenite (Donnerberg et al, 1989).

examined crystal. The first model explaining the mechanism of reduction of LN was proposed by Jorgensen and Bartlett, 1969:

$$\text{"LiNbO}_3\text{"} \rightarrow O_2 + V_O^{2+} + 4e^- \quad (66)$$

Assuming that the reduction process is accompanied not only by the loss of oxygen but also by the diffusion of Li_2O from the crystal, the reaction can be written in the form:

$$\text{"LiNbO}_3\text{"} \rightarrow O_2 + Li_2O + 3V_O^{2+} + 2V_{Li}^- + 4e^- \quad (67)$$

The second model, proposed by Smit (1984), includes Nb_{Li}^{4+} and is written in the following form:

$$\text{"LiNbO}_3\text{"} \rightarrow O_2 + Li_2O + 4Nb_{Li}^{4+} + 4e^- \quad (68)$$

Both of these contain conduction electrons. Energy considerations show that for any type of diffusion of Li_2O of the model (66) is less suitable than the model proposed by Smit (68). In the case of a LN with a shortage, the model (68) can be modified by formation of the internal defective structure of the LN, i.e. Li-vacancies inside ilmenite regions of the crystal. Donnerberg et al, (1989) proposed three models which depend on the extent of dissolution of ilmenite formed inside the crystal during the reduction process, expressed by the reactions (69), (70) and (71), respectively:

$$\text{"LiNbO}_3\text{"} + 2\left(Nb_{Li}^{4+} V_{Nb}^{5-}\right)^- \rightarrow 3/2\,O_2 + Nb_{Li}^{4+} + 2\left(Nb_{Li}^{4+} Li_{Nb}^{4-}\right) + 6e^{--} \quad (69)$$

$$"LiNbO_3"+2(Nb^{4+}_{Li}\ V^{5-}_{Nb})^{-}\ 3/2\ O_2 + Nb^{4+}_{Li}+(Nb^{4+}_{Li}\ Li^{4-}_{Nb}) +6e^{-} \quad (70)$$

$$"LiNbO_3"+2\left(Nb^{4+}_{Li}\ V^{5-}_{Nb}\right) \rightarrow 3/2O_2 + Nb^{4+}_{Li} + 6e^{-} \quad (71)$$

In these expressions, regions with lithium deficit are written as $(Nb^{4+}_{Li}V^{5-}_{Nb})$. Thus, the formation of electrons in the conduction band is a consequence of formation of ilmenite regions as a result of diffusion of lithium from these regions. Whilst in (69) the ilmenite structure remains stable during reduction, in reactions (70) and (71) it is seen to be at least partially dissolved. The calculated activation energy of the electrons is 1.7 eV for (69), 0.96 eV for (70) and 0.2 eV for (71). The authors have noted that all three examined versions are preferred in comparison to the model of oxygen vacancies (66). For example, the major difference between the reactions (69)–(71) and (66) is 0.6 eV per electron. Thus, on the basis of purely energy considerations it may be assumed that the ilmenite phase is most likely to form during the reduction process. It should also be mentioned that the reduction of lithium-deficit crystals is far easier than that of stoichiometric crystals.

19. Disruption of Stoichiometry in Lithium Niobate Crystals

19.1 Composition of LN crystals

The production of LN crystals with a homogeneous composition is an urgent task. The heterogeneity of chemical composition determines optical heterogeneity and also the scatter of other physical properties and, in particular, the speed of surface acoustic waves (SAW). Considerable advances have made in solving the problem of stabilising the composition along the length of the crystal boule of LN grown by the Czochralski method. However, investigations are being continued, as indicated by the appearance of new publications in this area.

To avoid disrupting the chemical composition of the crystal along the length of the boule, it is necessary to use a melt with a congruent composition which may greatly differ from stoichiometric $LiNb_3$. The congruent composition $\alpha(Li_2O + (1-\alpha) \rightarrow Nb_2O_5$, where $\alpha = 0.486$ (Byer, Young, Feigelson, 1970) has been determined more accurately quite recently: $\alpha = 0.4845$ (Grabmaier, et al., 1991) and $\alpha = 0.483$ (O'Bryan, et al., 1985). It should be mentioned that the congruent composition depends on the impurity composition of initial reagents and is determined, in our view, mainly by the disruption of stoichiometry of initial niobium pentoxide. The shortage of oxygen in niobium pentoxide (used for producing the starting melt) which is not taken into account disrupts the stoichiometry of lithium niobate crystals.

19.2 Defective structure of Nb_2O_5

It is well known that the niobium pentoxide can exist in 13 different crystallographic forms and has a superstructure (Kofstad, 1972). Three polymorphous modifications have been determined most reliably: low-temperature (γ-form), medium-temperature (β-form) and high-temperature (α-form) (Fairbrother, 1967). The β and α forms are structurally identical. The temperatures of the phase transformations from the amorphous to γ-phase (435°), from γ- to α-phase (830°) were determined. The densities of these phases at 25°C are equal to respectively 4.63, 5.17 and 4.55 g/cm^3 (Kofstad, 1972). The most stable high-temperature α-form belongs to the monoclinic crystal system, spatial group *P*2, the elementary cell contains 14 formula units. The cell parameters: $a = 21.16$, $b = 3.822$, $c = 19.35$ Å, β = –119° 50'. The complex crystal structure of niobium pentoxide leads to the formation of additional phases in the composition range from $Nb_2O_{4.6}$ to Nb_2O_5. The reflections of the niobium pentoxide on x-ray diffraction patterns appear at composition $Nb_2O_{4.78}$. The composition of niobium pentoxide, produced by the method of the chemical gas transport reaction, varies from $Nb_2O_{4.66}$ to Nb_2O_5. Thus, the niobium pentoxide may exist in a wide range of compositions Nb_2O_{5-x} which depends on its production technology. Figures 60 (a–c) shows x-ray diffraction patterns of niobium pentoxide produced from different sources (Fenske, 1988). The graphs show that the specimens greatly differ in their phase composition and, consequently, have different oxygen deficits

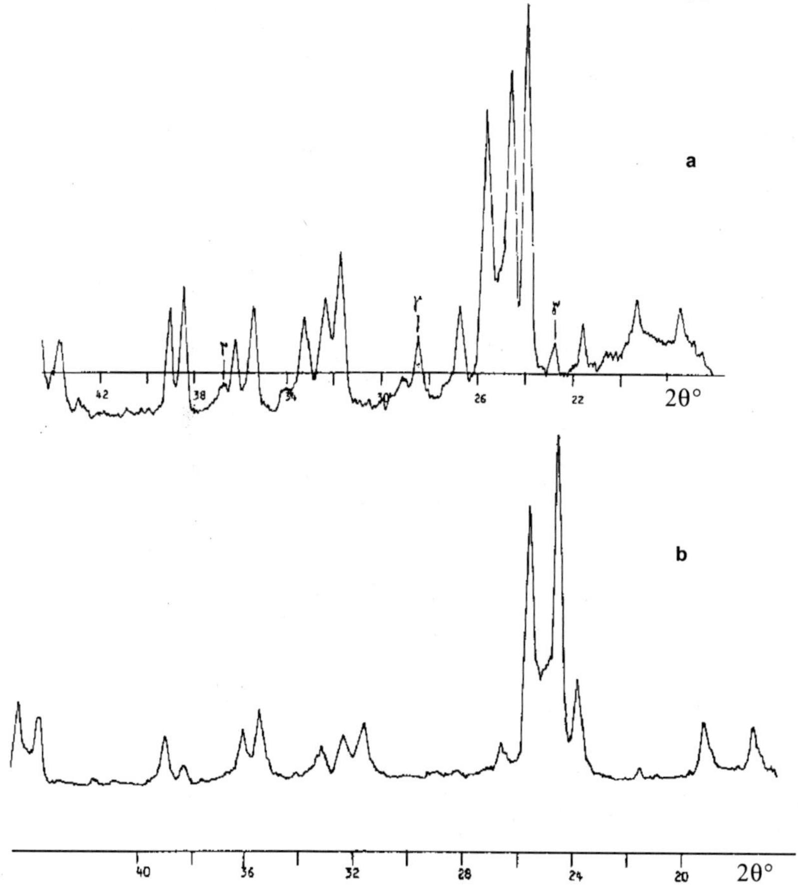

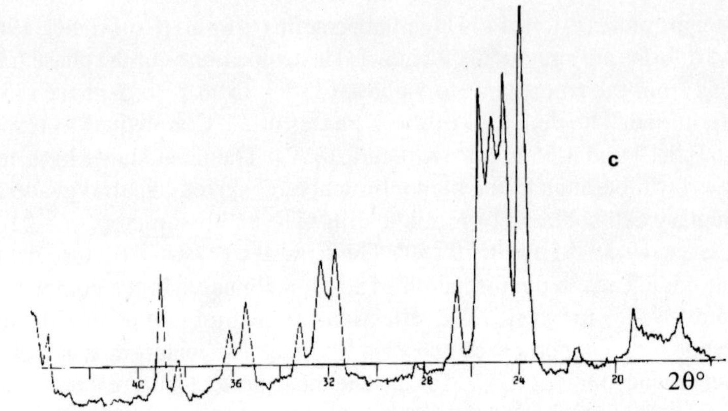

Fig.60 X-ray patterns of niobium pentoxide produced in: a) Germany (Stark pure grade) b); Hungary c); Russia (room temperature).

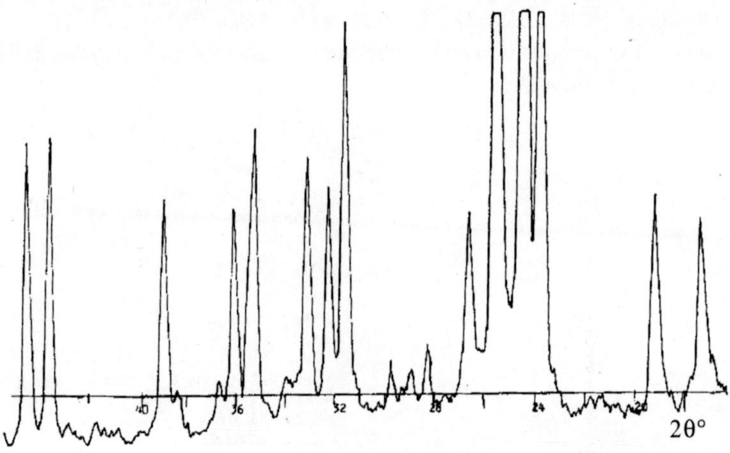

Fig.61 X-ray patterns of α-Nb_2O_5 (room temperature).

which must be taken into account in producing the starting melt for crystal growth. This leads to a recommendation for crystal technology: to stabilise the composition of Nb_2O_{5-x} at which x remains constant, it is necessary to transfer the niobium pentoxide to the most stable α-form. This is carried out by preliminary quenching the niobium pentoxide in air at 1200°C for 3 hours (Fenske, 1988). Fixing the composition Nb_2O_{5-x}, experiments were carried out to determine the congruent composition of the LN melt with special reference to the initial niobium pentoxide. The x-ray pattern of the α-phase Nb_2O_{5-x} is shown in Fig.61.

It was shown in subsequent investigations that the region of homogeneity of Nb_2O_{5-x} is restricted by the values of x from 0.100 (at 900°C) to 0.155 (1200°C). It was assumed that the degree of nonstoichiometry of the niobium pentoxide is determined by the simultaneous existence of anion oxygen vacancies and interstitial doubly-charged niobium ions.

Thus, the niobium pentoxide has an uncontrollable oxygen deficit. This is equivalent to a surplus of niobium.

19.3 Disruption of stoichiometry in lithium niobate crystals

It is was shown in Section 1 that the congruent composition of lithium niobate has a shortage of lithium oxide and a surplus of niobium oxide which substitutes lithium sites in the crystal lattice.

Lerner, et al. (1968) assumed for the first time that at a shortage of Li_2O in the lithium niobate crystals the niobium ions occupy lithium sites. Substitution takes place by the reaction:

$$6Li_{Li} + 3\,O_0 + Nb_{Nb} \rightarrow 3\,Li_2O + 4\,V_{Li}^{+} + Nb_{Li}^{4+}. \qquad (72)$$

Lerner assumed that the cells of the crystal lattice disappear from the crystal surface and niobium ions move into the bulk of the crystal and substitute the sites of absent Li^+. As a result of fulfilling the law of conservation of the charge the niobium ions assume the valency Nb^{4+}_{Li}.

Later, it was established that when reducing the lithium niobate crystals (annealing in a vacuum or an inert atmosphere at high temperatures) the density of the crystals increases. This increase is attributed (Smyth, 1983 and Schirmer, et al., 1991) to the substitution of lithium sites by niobium by the reaction:

$$LiNbO_3 + 2(V_{Nb}^{5-}\,Nb_{Li}^{4+}) \rightarrow 3/2O_2 + Li_{Li} + 2\,Nb_{Nb} + Nb_{Li}^{4+} + 6e \qquad (73)$$

where the $LiNbO_3$ molecules are distributed on the crystal surface. This model shows that oxygen leaves the surface of the crystal whereas Nb^{5+} forms complexes $(V_{Nb}Nb_{Li})$.

Examining the absorption spectra of lithium niobate crystals, Bollmann (1989) was forced to conclude that the models proposed described above cannot explain satisfactorily the absorption behaviour of the reduced crystals. The following reaction was proposed as a new model for explaining the observed dependence of conductivity and absorption in the crystal reduction stage of crystal reduction:

$$2\,Nb^{5+} + O^{2-} \rightarrow 2\,Nb^{4+} + V_O^{2+} + 2e^- + 1/2O_2 \qquad (74)$$

Free electrons are produced by ionisation $Nb^{4+} \rightarrow Nb^{5+} + e^-$.

Detailed examination of the defective structure of lithium niobate crystals, carried out by a group of Japanese scientists (Iyi, Kitamura, Izumi, et al., 1992) showed that substitution of lithium by niobium results in the formation of Li vacancies by equation (72). However, the variation of crystal density is described satisfactorily by both mechanisms (72) and (73).

In strong reduction of LN crystals at temperatures below 1000°C the weight loss is insignificant, i.e. the oxygen content of the crystal does not change. However, in growth of the crystal when the melt is known to contain a surplus of niobium, above the stoichiometry, free oxygen may form in the stage of formation of the crystal lattice with the formation of oxygen vacancies. The assumption according to which the substitution of lithium by niobium in the crystal lattice is accompanied by the formation of oxygen vacancies, will be taken into account in subsequent considerations. For practical purposes, it is interesting to determine the relationship between the disruption of stoichiometry in Nb_2O_{5-x} and the disruption of stoichiometry in the lithium niobate crystals α (Li_2O). The following procedure was used in calculations: molecular weights $M(Nb_2O_{5-x})$ were calculated and the difference $\Delta M = M(Nb_2O_5) - M(Nb_2O_{5-x})$ was determined. This was followed by determining the relative fraction of niobium $\Delta M_{Nb} = \Delta M.M(Nb_2)/M\ (Nb_2O_{5-x})$ and, finally, the atomic fraction of the surplus of niobium in niobium pentoxide $Nb_{2(1+y)}O_5$, where $y = \Delta M_{Nb}/A_{Nb}$, A_{Nb} is the atomic weight of niobium. The relationship between α and x is determined assuming that surplus Nb^{+5} ions in the lithium niobate lattice occupy most probably the Li^+ sites, with the formation of two neutral oxygen vacancies V_O. The structural formula of lithium niobate is written in the form:

$$\underbrace{[\alpha(Li_2O)+(1-2\alpha)(Nb_2O_{5-x})]}_{\text{Li-positions}}+\underbrace{[\alpha(Nb_2O_{5-x})]}_{\text{Nb-positions}} \qquad (75)$$

On the basis of the assumptions made above it follows that $2y = 1-2\alpha$ and, consequently, $\alpha = 0.5 - y$. The results of the calculations are presented in Table 7 and Fig.62 which show that $x = 0.13$ corresponds most accurately to the congruent composition $\alpha = 0.4845$. Thus, the structural formula of niobium pentoxide can be written in the form $Nb_2O_{5-0.13}$.

The formulae given above $y = 0.5-\alpha(Li_2O)$ makes it possible to calculate,

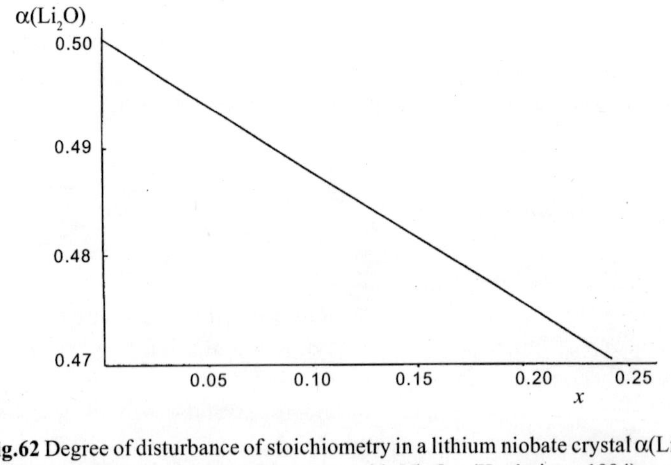

Fig.62 Degree of disturbance of stoichiometry in a lithium niobate crystal $\alpha(Li_2O)$ in relation to oxygen deficit x in the initial niobium pentoxide Nb_2O_{5-x} (Kuz'minov, 1994).

Table 7 Relationship between disturbance of stoichiometry in Nb_2O_{5-x} $\{Nb_{2(1+y)}O_5\}$ and the congruent composition of lithium niobate $\alpha(Li_2O)+Nb_2O_{5-x}$

X	M	ΔM	ΔM_{Nb}	y	α_L
0.00	265.82	0	0	0	0.5
0.05	265.02	0.8	0.561	0.006	0.494
0.06	264.86	0.96	0.673	0.0072	0.4927
0.07	264.7	1.12	0.786	0.0085	0.4915
0.08	264.54	1.28	0.899	0.0097	0.4903
0.09	264.38	1.44	1.012	0.0109	0.4891
0.10	264.22	1.60	1.125	0.0121	0.4879
0.11	264.06	1.76	1.238	0.0133	0.4868
0.12	263.9	1.92	1.352	0.0145	0.4854
0.13	263.74	2.08	1.456	0.0158	0.4842
0.14	263.58	2.24	1.579	0.0170	0.4830
0.15	263.42	2.40	1.693	0.0182	0.4817
0.20	262.62	3.20	2.264	0.0244	0.4756

for the known congruent point, the surplus of niobium in the melt and, vice versa, determine the congruent melt at the known surplus of niobium in the nonstoichiometric pentoxide. This is clearly seen in the Li_2O–Nb_2O_5 diagram (Fig.9) which shows the congruent composition.

19.4 Calculating the x-ray density of lithium niobate

To substantiate the main assumption according to which the stoichiometry of the initial niobate pentoxide leads to a deficit of lithium oxide in the congruent melt, the authors calculated the x-ray density of crystals for different $\alpha(Li_2O)$ and compared the results with the experimental values. Figure 63 shows two groups of experimental values of the densities of LN crystals (D_m) in relation to the degree of disruption of stoichiometry of $\alpha(Li_2O)$. These two groups greatly differ from each other but have a general tendency for a decrease of density on approaching total stoichiometry of $LiNbO_3$ (Grabmaier, 1989). To determine which group of the experimental values is closer to the true value, we calculated the x-ray density (D_x) using the precision parameters of the hexagonal lattice of the LN for different values of $\alpha(Li_2O)$ (Kovacs, Polgar, 1986; Abrahams, Marsh, 1986). The parameters a_H, c_H and the volumes of the elementary cells and x-ray densities calculated from them are presented in Table 8 and Fig.62. D_x was computed from the equation:

$$D_x = 11.5[\alpha\, M(Li_2O) + (1-\alpha)\, M(Nb_2O_5)] / a_H^2 c_H, \qquad (76)$$

where M are the molecular weights of lithium and niobium oxides. Of the calculated values of D_x only the density of the stoichiometric composition was close to the experimental value obtained by Grabmaier (1989), and the remaining values D_x greatly differed to higher densities. This discrepancy indicates that

Table 8 Lattice parameters, the volume of an elemantary cell and density of LN crystals in relation to disturbance of stoichiometry of the melt and the crystal

α_L	α_S	a_H, Å	c_H, Å	V, Å^3	D_x	D_A	D_m
0.470	0.4785	5.1514	13.872	318.884	4.639	4.653	–
0.475	0.4804	–	–	–	–	–	4.654
0.480	0.4818	5.1508	13.864	318.6257	4.649	4.655	4.659
0.4845	0.4828	5.1502	13.8636	318.5423	4.634	4.642	4.646
0.485	0.485	5.15052	13.86496	318.6132	4.646	–	–
0.486	–	–	–	–	–	–	4.643
0.490	0.4836	5.1494	13.863	318.4296	4.6477	4.639	4.655
0.499	–	–	–	–	–	–	4.636
0.500	0.4843	5.14739	13.8561	318.0235	4.649	4.631	4.634

D_x is the density of LN crystals, calculated with corrections for α_S and ΔD_x; D_A is the density calculated from cation distribution (Abrahams model); D_m is the experimental value of density

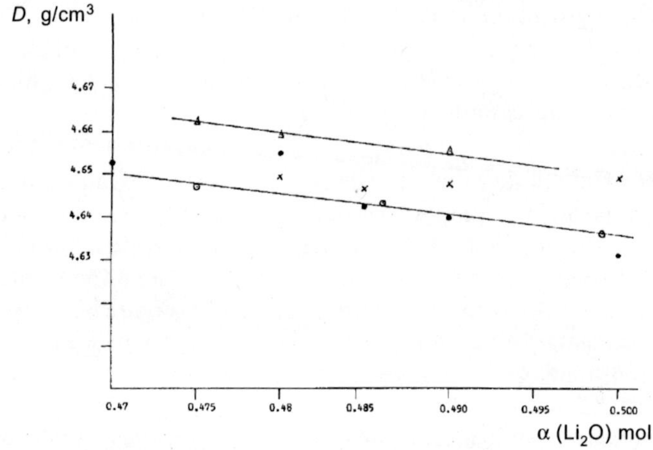

Fig.63 Density of lithium niobate crystals in relation to the degree of disturbance of stoichiometry of $\alpha(Li_2O)$ of the starting melt. ⊙) data obtained by Kovacs, Polgar, 1986; Δ data by Lerner, Legras, Dumas, 1968; ×) data obtained by the author of this book; ●) Abrahams, 1990.

actual crystals contain a large number of defects. For the congruent composition these defects may be represented only by oxygen vacancies since all metallic ions are included in the lattice of the growing crystal.

As mentioned previously, at a deficit of Li_2O the Nb^{5+} ions fill the lithium positions. To retain electroneutrality, two oxygen vacancies form for every Nb ion in the Li position. To match the computed and experimental values of the crystal density, a correction for the oxygen vacancy must be introduced into the calculated value. The corrections were calculated from the equation:

$$\Delta D_x = 184.004(1-\alpha)Z / a_H^2 c_H \quad (77)$$

The calculated density D_x is in best agreement with the experimental values in calculating the oxygen vacancies in the model in which

$$Z = \{4(1-2\alpha)+(1-\alpha)x\} \quad (78)$$

oxygen vacancies forms for every LN molecule. However, even this correction for the nonstoichiometry of niobium pentoxide does not make it possible to obtain densities close to experimental values. We have concluded that the disruptions of the stoichiometry of the crystalline LN $\alpha(Li_2O)$, published by Abrahams (1989), relate in fact to the starting melt. This is incorrect. The study by Grabmaier et al. (1990) contains an experimental relationship between the disruption of stoichiometry of the starting melt $\alpha_L(Li_2O)$ and the disruption of stoichiometry in the crystal $\alpha_S(Li_2O)$. On the basis of this dependence we determined the correspondence of the values of α_L and α_S. The x-ray density of the LN crystals was computed using the values of α_S corresponding to α_L and corrections for the disruption of stoichiometry of niobium pentoxide, $x = 0.13$. The resultant values were close to the experimental values obtained by Kovacs and Polgar (Kovacs, Polgar, 1986), and these results are presented in Table 8 and Fig.63.

Our approach has also been validated by calculations of D_x taking into account the experimentally determined distribution of Li and Nb cations in crystallographic sites by Abrahams and Marsh (Abrahams and Marsh 1988). For comparison, these values are also given in Table 8 and Fig.63. The fact that the experimental and calculated values of the densities of LN are similar confirms the model of distribution of the Nb cations in Li sites with the formation of oxygen vacancies. Thus, the congruent composition of the melt compensates the nonstoichiometry of niobium pentoxide and makes it possible to produce crystals with filled niobium and lithium sublattices. However, the crystals grown from the congruent melt contain point defects in the form of niobium atoms in the lithium sites and oxygen vacancies.

19.5 The Abrahams model

X-ray diffraction analysis of LN crystals of congruent composition, characterised by the Curie temperature $T_C = 1415$ K, showed that Li sites are occupied by lithium ions to 94.1 (3)% and 5.9 (3)% by niobium ions, Nb sites are occupied by niobium only to 95.3 (3)% and 4.7% of sites are vacant. Oxygen fully fills the crystal structure, i.e. 100%. The defective structure of LN of the congruent composition corresponds to the formula: [Li(0.941)Nb(0.059)] $Nb(0.953)O_3$.

To describe the distribution of cations in the crystallographic positions, Abrahams (1980) proposed a model in which missing lithium atoms are filled with niobium, and 4/5 of this number of vacancies forms in the niobium sublattice.

Oxygen is fully retained. The structural formula of LN in the general form is [Li(1–5u) Nb(5u)] Nb(1–4u) O_3. Lithium sites are enclosed in the square brackets. For the congruent composition u = 0.0118. Abrahams applied this model of distribution of the cations in the LN lattice to the composition range of $\alpha(Li_2O)$ from 0.5 to 0.47. Thus, the congruent composition of the melt contains a surplus of Nb atoms 0.059–0.047 = 0.012 mol. The structural formula of niobium pentoxide can be written in the form $Nb_{2+0.012}O_5$, i.e. y = 0.012 is close to the value which we determined previously, Table 7. Thus, the model proposed by Abrahams does not explain the origin of the surplus of niobium in lithium niobate and only states the fact of its presence in the crystals and, consequently, in the melt.

On the basis of these considerations it may be concluded that the surplus of niobium in the congruent melt is compensated by a shortage of the same amount of lithium, and oxygen vacancies form to retain electroneutrality. The niobium sublattice is completely filled. This is the difference in comparison with the model proposed by Abrahams. It should also be noted that the precision parameters of the lattice of the lithium niobate crystals of different stoichiometry, measured by Abrahams and Marsh, relate to the composition of melts from which crystals were grown. To calculate the x-ray density, it is necessary to introduce a correction for the difference in the compositions of the melts and crystals. From our viewpoint, the most perfect and chemically homogeneous lithium niobate crystals can be grown from a stoichiometric melt prepared from the stoichiometric niobium pentoxide Nb_2O_5. In this case, the stoichiometric composition of the melt coincides with the congruent composition. Consequently, problems arise in the technology of production of perfect lithium niobate crystals when producing the stoichiometric niobium pentoxide.

20 Effect of Isomorphous Impurities on the Stoichiometry of Lithium Niobate

When analysing the experimental data on the physical-chemical properties of crystals and melts of lithium niobate to determine the dependence of the physical properties of crystals on chemical composition, the author paid attention to large differences in some experimental data, mainly the dependences of crystal density and the speed of propagation of surface acoustic waves (SAW) on chemical composition (the lithium to niobium ratio). The principal difference between the dependences of the composition of the melt and the crystal growing from it was obtained by Carruthers et al (1971) and Grabmaier et al (1990). The Carruthers dependence indicates that at a specific composition of the melt it is possible to grow a crystal with stoichiometric composition, whereas the Grabmaier curve shows that only the crystals of congruent composition can be grown from the melt above the congruent point. The problem of producing stoichiometric crystals was solved by Japanese investigators by continuous feeding the melt in the crucible (Kitamura et al 1992). Attention should be given to the fact that the publications by Carruthers and Grabmaier were published with a gap of approximately 20 years. The large gap in publication separates the investigations on the basis of the density of crystals of various stoichiometry. It is possible that similar differences also exist for other crystal properties, for example, for the extraordinary refractive index. The dependence of the optical properties on composition has not been examined by the author. Here, we shall discuss these three special features.

20.1 Dependence of crystal density on chemical composition

The dependences of density on the lithium content of the $\alpha(Li_2O)$ crystal, obtained by Lener et al, 1968, on the one hand, and by Kovacs and Polgar (Kovacs, Polgar 1986), on the other hand, and also by Abrahams (Abrahams 1989), are shown in Fig. 63. The graph indicates that the linear dependences are parallel. The permissible errors are considerably smaller than the distance between them. The difference in the values of density (ΔD), determined at the congruent point $\alpha(Li_2O) = 0.485$ mol is $\Delta D = 0.014$ g/cm^3. It may be assumed that these dependences relate to different compositions: the compositions of crystals and the composition of melts from which the crystals have been grown. Unfortunately, the literature contains a large number of studies in which the authors do not take this large difference into account. In the examined case, these dependences would intersect at the congruent point at which the composition of the crystal corresponds to the composition of the melt. This has been observed in experiments.

Our assumption is reduced in the fact that in early stages of growing lithium niobate crystals technologists used niobium pentoxide containing isomorphous tantalum in the form of tantalum pentoxide. The uncontrolled presence of tantalum leads to an uncontrolled increase of crystal density. From the difference of the density $\Delta D(\alpha)$ of the crystals we can calculate the tantalum content of niobium and, if possible, compare it with the certified data of initial pentoxide available at that time.

In section 19, we proposed the mechanism of disruption of stoichiometry of lithium niobate crystals and presented a model of the crystal composition. The model is based on the fact that as a result of a shortage of oxygen in the initial niobium pentoxide a surplus of niobium appears in the melt. In the crystal, niobium occupies lithium sites, as confirmed by Abrahams and Marsh (Abrahams, Marsh 1986). The electroneutrality of the crystal is retained as a result of formation of oxygen vacancies. Some authors, including Abrahams, assume that oxygen fully fills its sites. However, in latest studies it was assumed (on the basis of energy calculations) that lithium niobate crystals contain fragments with an ilmenite structure which compensates the oxygen deficit (the crystal structure is eliminated, Bokii 1960). Taking this model into account, we proposed an equation for calculating x-ray density of a crystal in relation to the lithium content $\alpha(Li_2O)$ (equation 76).

To ensure that the densities calculated using this equation are in agreement with the experimental data, it is necessary to make corrections for oxygen vacancies and discrepancy of the crystal composition in relation to the melt composition according to the study by Grabmaier et al (1991). The last two corrections do not depend on the tantalum content of the initial niobium pentoxide and, consequently, have been ignored in this case. The molecular weight of niobium pentoxide, which contains tantalum pentoxide, is calculated from the equation:

$$M(\mathrm{Nb,Ta})=2[\beta A(\mathrm{Ta})+(1-\beta)A(\mathrm{Nb})]+5A(0), \tag{79}$$

where ß is the molar fraction of titanium, A are the atomic weights of niobium, tantalum and oxygen. The density difference is determined by the equation:

$$\Delta D_x = 11.5\ \{2(1-\alpha)\ \beta[A\ (\mathrm{Ta}) - (\mathrm{Nb})]\}/a^2_{\mathrm{H}}c_{\mathrm{H}} \tag{80}$$

The tantalum content is determined from the expression:

$$\beta = a^2_{\mathrm{H}}\ c_{\mathrm{H}}\ \Delta D_x/23(1-\alpha)[A(\mathrm{Ta})-A(\mathrm{Nb})]. \tag{81}$$

Substituting the values of constants for the congruent composition a_{H} = 5.15052, c_{H} = 13.86496, ΔD = 0.014 g/cm^3, α = 0.485, A(Ta) = 180.947, A(Nb) = 92.9064, ΔA = 88.0406, we obtain ß = 0.0051 moles = 0.51 mol% of tantalum in the initial niobium pentoxide. The calculated value is fully acceptable in the early stages of niobium production where there are still no stringent requirements on the removal of tantalum from niobium.

20.2 Dependence of the speed of surface acoustic waves on the chemical composition of the crystal

The situation, similar to the density considerations, is observed for the speed of propagation of surface active waves on the surface of crystal sheets of lithium niobate of the 128° XY cut. Japanese investigators found a linear dependence of the speed of propagation of surface acoustic waves on the lithium content of the crystal $\alpha(Li_2O)$ (Kushibiki et al 1991). A similar dependence was previously reported by other authors (Yamada et al 1987). Both dependences are shown in Fig.64. The dependences, constructed using the experimental points, are linear but do not coincide together and are not parallel. They also include comments according to which the difference is not associated with the systematic error or differences in the composition of the melt and the crystal. In the latter case, they would also intersect at the congruent point. If it is assumed that the difference in the speed of propagation of the surface acoustic waves is also caused by the presence of tantalum in initial niobium, then on the basis of the difference in the speed of propagation of the surface or acoustic waves it is possible to calculate the molar fraction of tantalum ß.

In the general case, the speed of propagation of an acoustic wave is determined by the equation (Farnell 1978):

$$V_l = (c/D)^{1/2} \tag{82}$$

where c is an elastic constant, D is the crystal density in previous symbols. We shall assume that the elasticity constant of the crystal does not depend strongly on the tantalum impurity and remains constant for the same direction of propagation of the SAW.

The speed equation shows that in the crystal relating to the straight line (2) in Fig. 29, the tantalum content is lower than for the straight line (1). To exclude the unknown elasticity constant c, we calculate the ratio of the speeds of propagation of the SAW at a fixed congruent point. To simplify calculations, the relationship will be written in the form:

$$\{1 + \Delta V/V_1\}^2 = \{1 + \Delta D/D_2\},$$

where ΔV is the difference of the speeds V_2 and V_1, D is the corresponding density difference. From the graphs we determine $V_2 = 3981$ m/s, $V_1 = 3925$ m/s, $\Delta V = 56$ m/s, the density at the congruent point $D_1 = 4.646$ g/cm^3. Calculations give $\Delta D = 0.1335$ g/cm^3, i.e. the value which is approximately an order magnitude higher than the value obtained in the previous section. It is possible that the speed of propagation of the SAW is also affected by other factors which have not been taken into account in these calculations. It should be noted that in practice the scatter of the speeds of the SAW in the sheets grown under plant conditions reaches 50 m/s which greatly complicate standardisation of the devices on the basis of SAW.

It is well known that when polishing crystalline lithium niobate with diamond paste the surface of the sheets contains an amorphous layer which then changes to a crystalline one with a greatly disrupted structure. The thickness of the disrupted layer can reach several micrometers and form a layer through which the SAW propagate. The density of the disrupted layer is slightly lower than the density of the undamaged part of the crystal. It is possible that this reason explains the given large difference in the speeds of propagation of the SAW. However, in this case, the method of determining the composition on the basis of the speed of the SAW, proposed by Kushibiki et al, 1991, can be applied only to the reproducible technology of polishing sheets which is difficult to achieve in practice. This element of indeterminacy cast doubts on the proposal of the authors.

The propagation of the SAW through the crystalline sheet is also characterised by the absorption coefficient which is influenced by a large number factors, including: polishing, adsorption of moisture and gases, trivial contamination, the loss of oxygen during the process of polarization and other factors (Lewis, West, 1990). If the factors of the type of trivial contamination are excluded, the controlling factor will be the disruption of the surface layer, as indicated by the equation of the absorption factor:

$$\gamma = w^2 \ \vartheta_{eff}/2DV^3 \qquad (83)$$

where w is ultrasonic frequency, ϑ is the viscosity factor, other symbols are as previously. Thus, the absorption factor γ can be used to standardise the treatment conditions of the sheet surface. The method with these considerable restrictions can be recommended for determining the crystal composition.

20.3 Dependence of crystal composition on the composition of the melt from which the crystal is grown

In this section, we shall discuss the experimental dependences presented in the studies by Carruthers (Carruthers et al 1991) and Grabmaier (Grabmaier et al 1991) which are reproduced in Fig.8 using the same scale. The Grabmaier data were used to make corrections in calculating the x-ray density of lithium niobate in relation to the crystal composition and enabled the calculated data to be matched quite efficiently with the experimental values (section 19). Analysis of the Grabmaier dependence in Fig.8 results in a paradoxic dependence from the viewpoint of the technologist: when the composition of the melt is higher than the congruent point the crystals have congruent composition. In fact, the Grabmaier dependence reflects the fact that the melt contain a surplus of niobium as a result of a shortage of oxygen in the initial pentoxide. Thus, a structure with fully filled cation sublattices but with oxygen vacancies forms. At high temperature, close to melting point, the oxygen vacancies are ordered, i.e. the oxygen plane disappears and a fragment of the structure of ilmenite appears. As indicated by the phase diagram proposed by Reisman and Holtzberg (Reisman, Holtzberg 1958) with decreasing temperature the field of existence of the lithium niobate structure is greatly reduced in size to the composition close to congruent, and the distribution factor of lithium decreases with increasing lithium content of the melt from 1.02 to 0.98 when approaching the congruent composition (O'Bryan, Gallagher, et al, 1985). When the Li composition of the melt is higher than the congruent composition, the surplus lithium plays the role of an impurity and can form growth bands at fluctuations of temperature at the solidification front.

For the dependence published by Carruthers et al, Fig. 8, these considerations required further explanation because of the possible presence, in the initial niobium pentoxide, of the impurity of tantalum pentoxide which has already been discussed. In the melt volumes of the same weight the content of tantalum ions is approximately half the niobium content. Thus, surplus niobium is compensated by the presence of tantalum which results in the formation of vacant sites for niobium. The range in which niobium compensation takes place can extend outside the stoichiometric composition range: all depends on the shortage of oxygen and the tantalum content of the initial niobium pentoxide. The phase diagram, plotted by Reisman and Holzberg, has a wide range of existence of lithium niobate at high temperatures. In this situation, is possible to grow a crystal with the composition close to stoichiometric, as also indicated by the Carruthers dependence. The above considerations make it possible to explain qualitatively the difference between the two dependences examined.

Thus, the shortage of oxygen in the oxides of metals and the presence of isomorphous elements in the initial reagents which are often ignored by technologists form boundaries in the production of crystals of high optical quality and with reproducible physical properties, as is the case with lithium niobate crystals. It should be noted that the tantalum content of the powder of niobium pentoxide and lithium niobate of the highest qualification exceeds the content of other elements by two orders of magnitude. An example of other isomorphous elements can be rare-earth elements whose oxides are usually contaminated by their analogues.

21. Growth and Characterisation of Lithium Niobate Crystals Doped with Magnesium

Zong et al. (1980) reported that the resistance of lithium niobate crystals to optical damage greatly improves if more than 4.5 mol% MgO is added to the melt from which the crystal is grown. The $MgO–Nb_2O_5–Li_2O$ ternary phase diagram was investigated for the first time by Grabmaier et al. (1986). They established that there is a wide range of the solid solution in the $MgO–LiNbO_3$ pseudosystem extending to more than 25 mol% MgO. Usually, magnesium is added to the melt with a congruent composition, but at this high MgO concentration the congruent composition itself should change. The distribution coefficient (K_c) of MgO in the lithium niobate crystals, determined by Grabmaier and Otto, is 1.2, irrespective of the doping level. However, Furukawa et al. (1990) found that the MgO concentration gradually decreased along the growth direction and also decreased linearly with the doping level. When the MgO content is reduced from 5 to 3 mol.%, the distribution coefficient decreases from 1.2 to 1.04, respectively. The distribution coefficient K_c was determined more accurately in other studies (Zhon, et al., 1991).

The technology of growing lithium niobate crystals doped with magnesium differs only slightly from that used for undoped crystals, but the drawing rates of the crystals are usually considerably lower. The melt for growing crystals was produced as follows: chemical agents with high purity were mixed in alcohol in an agate mill. The mixture was compacted into discs which were then annealed for 6 hours at 900°C. The annealed specimens were again ground and compacted. Final annealing was carried out in air at 1150°C for 9 hours to ensure that the reaction takes its course.

The crystals doped in magnesium were grown by the Czochralski method, using automatic control of the crystal diameter on the basis of the weight of the crucible and the melt. Crystals were grown in the direction of the *Z* axis, the composition of the melt was Li/Nb = 47.2/52.8 with two doping levels of 5 and 7 mol.% MgO. The temperature gradient above the melt surface was 20–30°C/cm. Other crystal growth parameters: drawing speed 2.5 mm/h, rotation speed 20–22 rpm, cooling rate around 50°C/h. The grown crystals were annealed in an isothermal zone at 1150°C for 12 h and polarised at 1170°C for 40 min in a constant field at a current density of 5–8 mA/cm^2. The crystals were slowly cooled to a temperature approximately 100°C below the Curie point and the field was then removed.

The experimental results show that the melting point varies in relation to doping with MgO, at a constant ratio of the main components, and reaches the maximum value at 5 mol.% MgO and then decreases with decreasing MgO. At a constant MgO content the melting point of the crystals has a tendency to increase with a decrease of the Li/Nb ratio. Figures 64 and 65 show that the crystals containing 5–6 mol.% MgO and Li/Nb = 47.2/52.8 have the highest melting point and, consequently, the congruent composition of the LN, doped with magnesium, is close to this composition. Table 9 gives the effective distribution coefficients K_c and variations of the composition of lithium niobate crystals, doped with magnesium. The table shows that the variation of the Li_2O content from the upper to lower parts of the crystal, grown from a melt with a composition Li/Nb = 47.2/52.8, is very small and K_c (Li_2O) is approximately 1.01, which is close to values determined by Grabmaier and Otto (1986). The crystals grown from the melt with the composition Li/Nb = 47.2/52.8 and 5 mol.% MgO have a highly homogeneous composition and good optical quality. The crystals with the same ratio of the components but with 7 mol.% MgO were of lower quality. Thus, it may be concluded that at 5 mol.% MgO the congruent composition of the melt corresponds to Li/Nb = 47.2/52.8. However, it should be mentioned that doping the lithium niobate with magnesium 'erodes' the range of existence of the congruent composition. The dependence of the MgO content in the crystal on the content in the melt is shown in Fig.66 (Hy, Chang, 1991). The graph shows that the distribution coefficient K_c for concentrations

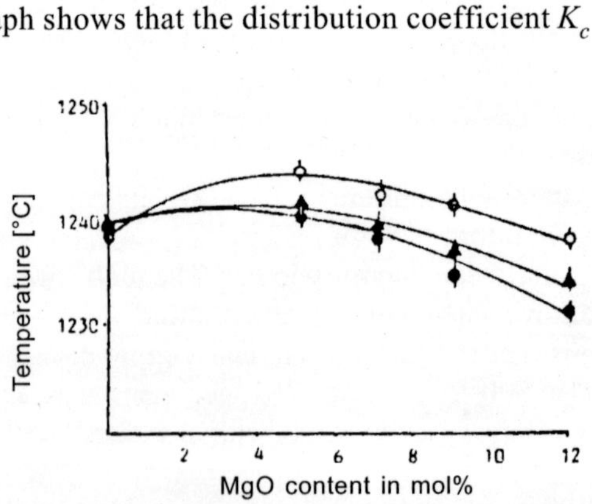

Fig.64 Melting temperature of MgO-doped $LiNbO_3$ versus MgO content of samples: (○) Li/Nb = 47.5/52.5; (●) Li/Nb = 48.2/51.8; (▲) Li/Nb = 48.6/51.4 (Zhou, Wang et al, 1991)

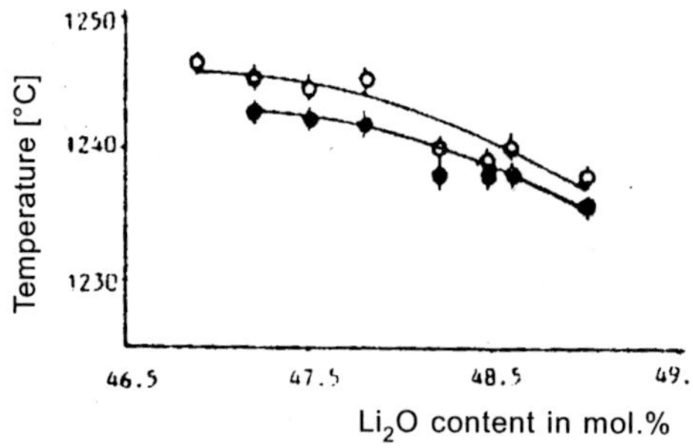

Fig.65 Melting temperature of MgO-doped $LiNbO_3$ versus Li/Nb ratio of samples: (○) MgO = 5 mol%; (●) MgO = 7 mol% (Zhou, Wang et al 1991).

not exceeding 5–7 mol.% reaches 1.2 and for higher concentrations decreases to 0.95. A two-phase system forms at more than 25 mol.% MgO in the melt. Figure 67 shows powder x-ray diffraction patterns of crystals containing 25 and 30 mol.% MgO. At 30 mol.% MgO, in addition to the LN lines, the x-ray diffraction pattern contains in all likelihood lines of $MgNb_2O_6$ (Hu, Chang, 1991).

Furukawa, et al. (1990) grew lithium niobate crystals doped with 1.3 and 5 mol.% MgO two inches in diameter. The drawing rate was varied from 1 to 3 mm/h, rotation speed 8–20 rpm. The temperature gradient was varied in the range 20–50°C by regulating the position of the crucible with the melt. Thus, the growth conditions greatly change with increasing crystal diameter. To obtain a monodomain structure throughout the entire doped crystal, the polarization temperature should be slightly higher than the Curie point. The field, applied to the crystal, also increases depending on the magnesia content. The polarization conditions for the pure and doped crystals are presented in Table 10. Attention should be given to the fact that an increase of the magnesium concentration results in difficulties in producing the monodomain state caused possibly by the formation of traps at domain boundaries or by a reduction of the strength of the effective field as a result of increasing electrical conductivity. The dependence of the Curie temperature on the MgO content is shown in Fig.68. The domain structure of the polydomain crystal of the *Z*-cut has the form of concentric rings

Table 9 Distribution factor K_c and the variation of the composition of LN single crystals, doped with MgO, with Li/Nb = 47.2/52.8 (Zhong et al, 1980)

Specimen	Components	Melt (vol %)	Crystal		K_C
			Top	Bottom	
L N M–1	MgO	5.00	5.36	5.20	1.07
	Nb_2O_5	50.16	49.36	49.48	0.98
	Li_2O	44.84	45.25	45.26	1.01
L N M–2	MgO	7.00	6.72	6.65	0.96
	Nb_2O_5	49.10	48.77	49.07	0.99
	Li_2O	43.90	44.51	44.28	1.07

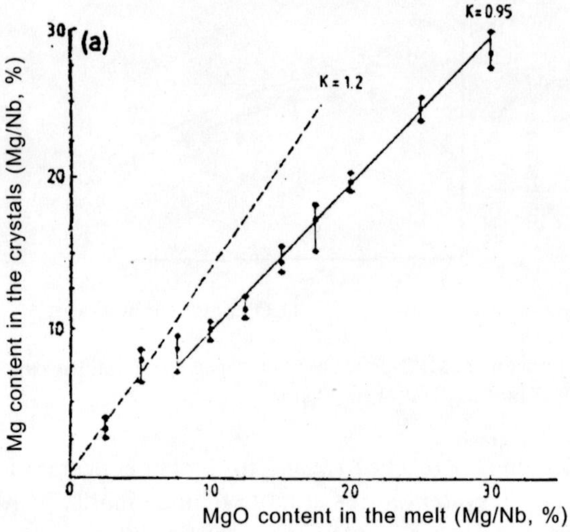

Fig.66 Dependence of the Mg^{2+} content in the crystal on the content in the melt (Hu L.J., Chang Y.H., 1991).

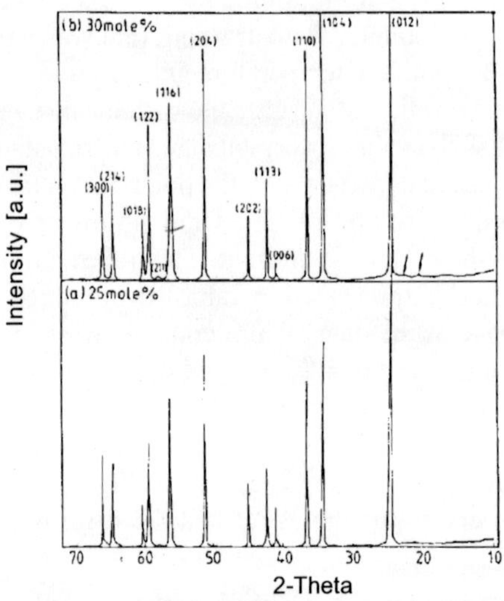

Fig.67 Powder X-ray patterns of pure lithium niobate crystals and of lithium niobate crystals doped with MgO (Hy L.J., Chang Y.H., 1991).

typical of the undoped crystal. These rings correspond to the bands of intersection of the phase boundary by the plane. Examination of polarised crystals, doped with 5 mol.% MgO, by x-ray topography showed individual fragments of a domain structure indicating that the polarization process was incomplete (Furukawa 1990).

The authors also noted an increase of the size of the intergranular boundary with increasing magnesia content of the crystal. For example, the lower half

Table 10 Polarization conditions on undoped LN crystals and LN crystals doped with MgO (Furukawa et al, 1990)

Specimen	Curie temperature T_C (°C)	Applied field (V/cm)	Result
Undoped LN	1150	2.0	good
Doped with MgO, 1 mol %	1170	2.0	poor
	1170	3.5	partly poor
	1170	7.0	good

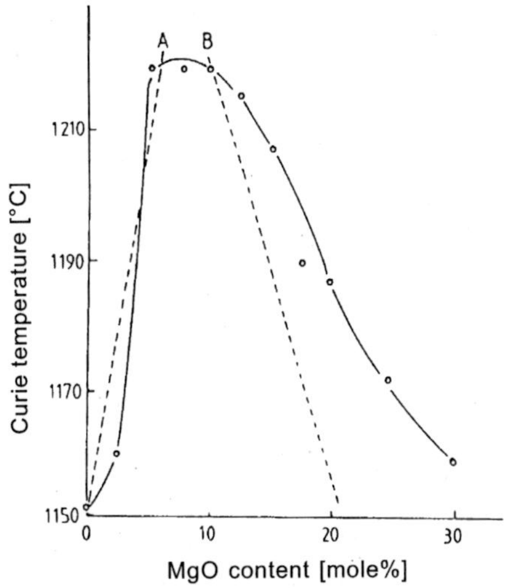

Fig.68 Dependence of the Curie temperature on the MgO content of lithium niobate crystals (Hu, Chang, Yen, 1991).

of the crystal, doped with 3–5 mol.% MgO, contained clusters of intergranular boundaries with a higher content of magnesium ions.

Determination of the maximum concentration of MgO and also the determination of the congruent composition are important for the technology of these crystals. For this reason, the dependence of the lattice parameters on the magnesia content of crystals was studied (Hy, Chang, 1991). The lattice parameters increase when the magnesium concentration is increased to approximately 25 mol.%. A further increase to 30 mol.% leads to an unexpected decrease of both lattice parameters, Fig.69. Stress relation of the lattice possibly leads to the precipitation of a second phase, as indicated by the diffraction peak of the specimens with 30 mol.% MgO (Fig.67). Thus, the solubility limit of MgO in lithium niobate is between 25 and 30 mol.%. This is also indicated by the dependence of the transmission factor on doping: the crystal becomes non-transparent on approaching

30 mol.% MgO. The authors also examined the dependence of density and Curie temperature on the MgO content of the melt. These dependencies show considerable changes if the addition concentration exceeds 7.5 mol.%. Crystal density rapidly decreases for highly doped materials. The Curie temperature has a broad maximum in the concentration range 5–19 mol.% MgO. A decrease of Curie temperature for highly doped materials which coincides with the decrease of the density is probably associated with the formation of Li vacancies when Li^+ is substituted by Mg^{2+}.

Examining specimens by nuclear magnetic resonance, Peterson and Carnevale (1972) established that 8% of niobium ions in undoped LN crystals occupy lithium sites. This distribution was confirmed much later by Abrahams and Marsh (1988) by x-ray diffraction analysis. The reason for this distribution of the cations was explained in the previous section as the result of disruption of the stoichiometry of the initial niobium pentoxide. Doping the lithium niobate with magnesium reduces the concentration of Nb ions in Li sites. In the specimens containing more than 7.5 mol.% MgO Nb is not found at all in the Li sites. Table 11 shows the distribution of ions in the structure of the lithium niobate crystals doped with 9 and 20 mol.% MgO (Hy, Chang, 1991). Examining Table 11, it can be concluded that implantation of Mg cations into LN crystals reduces the number of niobium ions in A-sites and removes vacancies in B-sites which were observed by Abrahams and Marsh. On the other hand, the implanted Mg cations occupy A-sites and prevent surplus Nb cations from occupying the same sites, if the MgO concentration is greater than 5 mol.%. In addition, Mg additions are added up, the number of Li cations in the A-sites decreases and the vacancies in the A-sites increases. This is in agreement with a reduction of the crystal density and its Curie temperature. The variation of the distribution co-

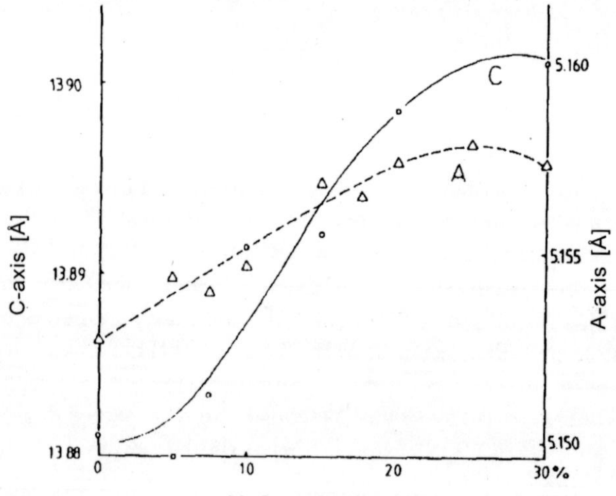

Fig.69 Dependence of the lattice parameters of lithium niobate on the MgO content of the crystal (Hu, Chang, Yen, 1991).

Table 11 Distribution of atoms in the crystal lattice in undoped lithium niobate and lithium niobate doped with MgO (mol%)

Positions	Undoped	9 mol% MgO	20 mol% MgO
		in melt	
Li	94.1 Li	91.0 Li	70.0 Li
	5.9 Nb	0.0 Nb	15.0 vacauies
		9.0 MgO	15.0 Mg
Nb	95.3 Nb	96.3 Nb	97.0 Nb
	4.7 vacancies	3.7 vacancies	3.0 vacancies
0	100.00	100.00	100.00

efficient from 1.2 to 0.95 may be associated with a change in the mechanism of implantation of the Mg cations in A-sites at MgO concentrations higher than 7.5 mol.%. This substitution is unfavourable from the energy viewpoint.

22. Doping Lithium Niobate Crystals with Titanium Oxide

Doping of lithium niobate with titanium oxide TiO_2 also leads to the redistribution of ions in the crystal structure of lithium niobate (Gallagher, O'Bryan, 1988). If it is assumed that after adding TiO_2 the anion sublattice remains completely filled, and the titanium ions, like magnesium ions, initially fill A-sites, we can write an equation for the concentration of resultant vacancies for different stoichiometry of the initial lithium niobate and the TiO_2 content. For the single-phase $Li_y\,Nb_{1-y}Ti_zO_{2.5-2y+2z}$ system, the fraction of vacancies in B-sites is:

$$C_V = (2 - 4y + z)/3 \qquad (84)$$

The calculations do not distinguish between the cation vacancies induced by a deficit of lithium ions or a surplus of magnesium and titanium. For the most interesting partial cases, these dependences are described by the equations:

$$C_V = 0.6667 - 1.3333y \ (z = 0) \qquad (85)$$

$$C_V = 0.3333z \ (y = 0.5000, \text{stoichiometry}) \qquad (86)$$

$$C_V = 0.02067 + 0.3333z \ (y = 0.4845, \text{ congruent}) \qquad (87)$$

Using the dependence of Curie temperature T_C on the lithium content of the crystal (or in the melt) (Guenbais, et al. 1981) (see also Fig.25), we can plot the dependence of T_C as a function of the fraction of cation vacancies C_V, as shown in Fig.70. The linear dependence of T_C on the content of cation vacancies or a lithium deficit can be approximated by the expression

$$T_C(°C) = 1204 - 3280\,C_V. \tag{88}$$

The experimental values of T_C, determined by the DTA method (Gallagher, O'Bryan, 1988) and also the calculated concentrations of the cation vacancies are shown in Table 12 and Fig.70. The experimental values of T_C are in good agreement with the calculated values, with the exception of high TiO_2 concentrations. The limiting TiO_2 concentration for the congruent melt is $C(TiO_2) = 0.20$.

The limiting solubility of titanium oxide in lithium niobate corresponds to the fraction of cation vacancies $C_V = 0.074$ and $T_C = 963°C$ and is equivalent to the value $y = 0.445$ which is in good agreement with the lithium content at the $LiNbO_3$–$LiNb_3O_8$ phase boundary at 1100°C (Svaasand, et al, 1974). A similar phenomenon is observed when magnesium oxide is added to the lithium niobate melt.

23. Physico-Chemical Properties of Niobium Pentoxide

Niobium pentoxide Nb_2O_5 is the main component forming the crystal lattice of lithium niobate. The crystals contain more than 90 wt.% of this component and the pentoxide has therefore a controlling effect on the physico-chemical properties of lithium niobate crystals. The physico-chemical properties of niobium pentoxide have already been partially examined in section 19. Here we shall attempt to show that the disruption of stoichiometric niobium pentoxide is a common property of metal oxides.

The degree of nonstoichiometry and the concentration of defects in the oxides

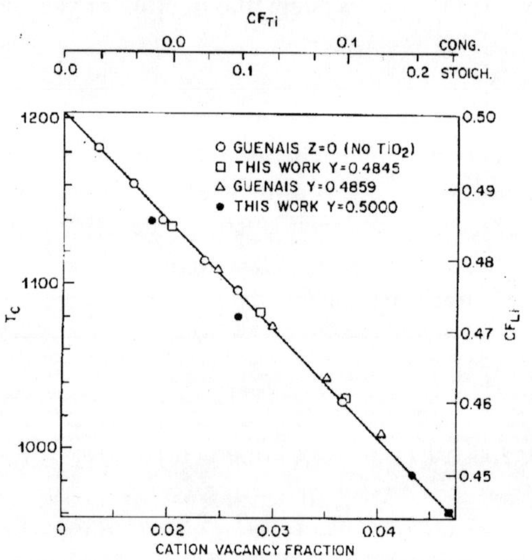

Fig.70 Plot of T_C, cation fraction of Ti added to either stoichiometric or congruent $LiNbO_3$ and cation fraction of Li as a function of the cation vacancy fraction for $LiNbO_3$ (Gallagher, O'Bryan, 1988).

Table 12 Curie temperature in relation to the composition of $Li_yNb_{1-y}Ti_zO_{2.5-2y+2z}$ compound (Gallagher et al, 1988)

$C_{Ti}(z)$	T_c (determined, °C)	C^*_V
	Stoichiometry (y = 0.500)	
0.00	–	0.000
0.05	1129	0.020
0.10	1078	0.038
0.20	988	0.066
	Congruent (y = 0.04845)	
0.00	1137	0.020
0.05	1084	0.037
0.10	1034	0.052
0.20	963	0.074

C^*_V is the quantity corresponding to Curie temperature T_C= 963°C in equation (88)

depend on the temperature and partial pressure of components. For oxides with a deficit of oxygen, the reaction leading to the appearance of nonstoichiometry can be written in the following general form:

$$\mathrm{M\ O} \rightarrow \mathrm{M\ O}_{1-x} + x/2\,\mathrm{O}_2. \quad (89)$$

From equation (89) we can conclude that an oxygen deficit increases with a decrease of the partial oxygen pressure because the partial pressure of metal vapours is negligible in comparison with the partial oxygen pressure. The nonstoichiometry of oxides is determined by the interaction with the oxygen of the atmosphere surrounding the oxide.

Oxygen vacancies form in transfer of oxygen atoms from lattice sites to the gas medium. The number of sites does not change. Assuming that the oxygen vacancy is neutral, this reaction can be written in the form

$$\mathrm{O_O} \rightarrow V_\mathrm{O}^* + 1/2\,\mathrm{O}_2. \quad (90)$$

The relationship of two electrons with the neutral oxygen vacancy can be disrupted as a result of thermal excitation. The oxygen vacancy, acting as a donor, becomes singly- or doubly-charged:

$$V_\mathrm{O}^* \rightarrow V_\mathrm{O}^+ + e^-,\ V_\mathrm{O}^+ \rightarrow V_\mathrm{O}^{2+} + e^-. \quad (91)$$

Free electrons are localised on cations in corresponding lattice sites. For oxides, M O reactions are written in the form

$$M_M + V_O^* \rightarrow V_O^+ + M_M^-, \; M_M + V_O^+ \rightarrow V_O^{2+} + M_M^-. \tag{92}$$

In these processes, the valency of the ion M in the crystal decreases to unity.

The deficit of oxygen in the oxides in relation to the stoichiometric composition is equivalent to a surplus of metal at which interstitial ions M_i^* are the main type of defect. Formation of an interstitial ion is accompanied by the transfer of atoms O_O to the gas phase by the reaction

$$M_M + O_O \rightarrow M_i^* + 1/2O_2. \tag{93}$$

As a result of this reaction, the number of cation and anion sites decreases to unity. Ionisation of atoms M_i^* leads to the formation of singly- or doubly-charged interstitial ions and free electrons.

The equilibrium of defects in crystals can be described using the law of acting masses, as in the case of ordinary chemical reactions. In oxides with a deficit of oxygen in which oxygen vacancies and electrons prevail, the electron concentration depends on the pressure of oxygen in the surrounding atmosphere. The electron concentration increases with decreasing oxygen pressure in accordance with the law varying from $p_0^{-1/4}$ to $p_0^{-1/6}$. Consequently, the electroneutrality condition can be approximated by the equality $2\,[V_O^{2+}] = n$, where n is the number of electrons.

Taking into account the law of acting masses, we can determine the concentration of oxygen vacancies in limiting cases (Kofstad, 1972):

$$\left[V_O^*\right] >> \left[V_O^+\right] + \left[V_O^{2+}\right], \left[V_O\right]_{tot} = K_1\, p_{O_2}^{-1/2} \tag{94}$$

$$\left[V_O^{2+}\right] >> \left[V_O^*\right] + \left[V_O^+\right], \left[V_O\right]_{tot} = \left(1/4K_1K_aK_b\right)^{1/3} p_{O_2}^{-1/6} \tag{95}$$

$$\left[V_O^{2+}\right] >> \left[V_O^*\right] + \left[V_O^+\right], \left[V_O\right]_{tot} = \left(1/4K_1K_aK_b\right)^{1/3} p_{O_2}^{-1/6} \tag{96}$$

Thus, for oxides with an oxygen deficit the dependence of the concentration of oxygen vacancies on the oxygen pressure is governed by the law varying from $p_{O_2}^{-1/2}$ to $p_{O_2}^{-1/6}$. In the oxides, formation of doubly-charged oxygen vacancies at high temperatures is probably more likely than the formation of neutral or singly-charged vacancies.

For oxides with an oxygen deficit, it is difficult to say in advance whether oxygen vacancies or interstitial ions of the metal will prevail. In the normal range of deviation from stoichiometry defects of both types are important. At reduced temperatures, many refractory oxides are stable even at low oxygen pressure. Therefore, oxygen vacancies may prevail at high partial oxygen pressures and small deviations from stoichiometry, and the interstitial cations are the main type of defect at low partial oxygen pressures and relatively large deviations from stoichiometry.

The equation describing the fraction of vacancies in the solid (or the equilibrium constant for the process of their formation) includes the vibrational entropy:

$$K = \exp(-\Delta G_f / RT) = \exp(-\Delta S_f / R)\exp(-\Delta H_f / R), \tag{97}$$

where ΔG_f is the variation of the free molar energy, ΔS_f is the variation of vibrational entropy, ΔH_f is the variation of vibrational enthalpy related to 1 mole. For niobium pentoxide $\Delta H_{VO^{\cdot\cdot}}$ = 102 kcal/mol, and the entropy of formation of doubly-charged oxygen vacancies for Nb_2O_5 is estimated to be 15 entropy units which is equal to 15 × 4.19 = 62.85 kcal/mol K.

If defects form as a result of interaction with gaseous oxygen, the entropy of their formation also includes the entropy of gaseous oxygen (S_{O_2}). The equation describing the formation of oxygen vacancies and localisation of electrons on cation sites is written in the form:

$$2\mathrm{Nb_{Nb}} + \mathrm{O_O} = V_{\mathrm{O}}^{2+} + 2\mathrm{Nb_{Nb}^{-}} + 1/2\mathrm{O_2}\,(\Delta S_{V^{2+}} = 15 \text{ entropy units}) \tag{98}$$

In this case, Nb_{Nb} and Nb^{-}_{Nb} denote the ions of pentavalent and tetravalent niobium, respectively, located in the lattice site. The dissociation entropy of Nb_2O_5 to NbO_2 at which the valency of niobium changes to 4 with simultaneous release of half of the oxygen mole, is 17.7±3 entropy units.

Previously (paragraph 19) it was mentioned that the structures of different polymorphous modifications of niobium pentoxide are very complicated. The high-temperature modification α-Nb_2O_5 has the range of compositions (NbO_x) $2.4 < x < 2.5$ which is associated with oxygen loss in the vicinity of the melting point. However, Magneli identified two previously unknown compounds with the formulae $NbO_{2.46}$ and $NbO_{2.40}$ existing in the temperature range close to the melting point which can be regarded as stable reduced niobium oxides with the general formula $B_{3n}O_{8n-3}$. At $n = 3$, the formula corresponds to B_3O_7 ($BO_{2.333}$) and at $n = 4$ to $B_{12}O_{29}$ ($BO_{2.417}$). The term of this homological series $B_{15}O_{37}$ ($BO_{2.467}$) at $n = 5$ could not be obtained, but the reduced oxide has a composition corresponding to this homologue.

The structure of these homologues can be described by the combination of two structural elements shown in Fig.71 (Mandelcorn, 1964). Element D is a binary series of closely linked octahedrons and E is a single series found in both compounds (see Fig.71). The elements D and E are connected by the tips of the octahedrons. The structure of the phase B_3O_7 ($n = 3$), presented in the shortened form is:

– D – E – D – E–...

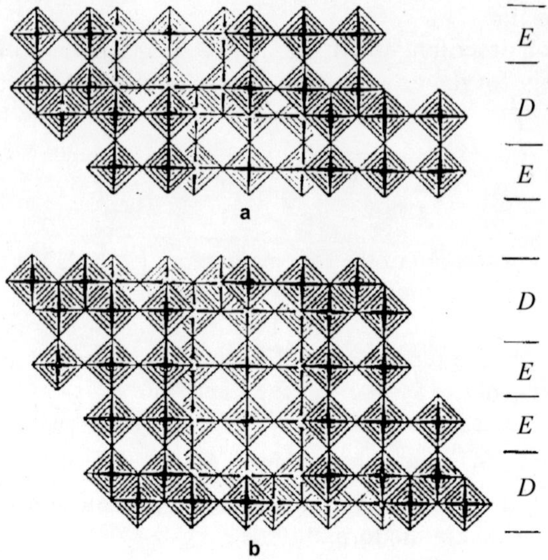

Fig.71 Structures of $TiNb_2O_7$ (a) and $Ti_2Nb_{10}O_{29}$ (b): a) consists of alternating cells of D and E; b) formed by alteration of cells D with two cells E (Mandelcorn L., Edited, 1964)

and the structure of the phase $B_{12}O_{29}$ ($n = 4$) can be written in the form:

...– D – E – E – D – E – E – D –...

A special feature of this group of oxides is the regular alternation of the structural elements or dimensions of the blocks of the phase. The literature contains no reports on compounds intermediate between the adjacent terms of a homological series containing structural fragments of both phases in different sequences.

Thus, the variations of the oxygen content of the initial niobium pentoxide can be partially explained by random changes of the number of elements D and E. If these fragments do not extend throughout the entire crystal, this may result in the formation of holes in the frame lattice of oxygen atoms. Possibly, a similar phenomenon is also observed in the formation of the crystal lattice of lithium niobate during crystal growth.

24. Growth of Lithium Niobate Crystals by Czochralski method

24.1 Preparatory operations for crystal growth

Preparation of the charge

The components Li_2CO_3 and Nb_2O_5 are of the special purity grade. If necessary, the reagents are milled and screened. Screening is essential for removing secondary objects which are sometimes found in the reagents. To remove moisture, Li_2CO_3 is dried at a constant rate at 300°C for 3–5 h. Drying is carried out in a normal drying cabinet. To determine the congruent composition and its reproducibility, initial Nb_2O_5 should be transferred to a stable monoclinic phase (α-phase). For this purpose, niobium pentoxide is annealed in a muffle furnace in air at 1000–1200°C for 3–5 h. The annealing temperature and time are selected by experiments to ensure complete transfer of Nb_2O_5 to the monoclinic phase. The extent of the reaction is inspected by x-ray phase analysis. We have already discussed the variable composition of niobium pentoxide and its effect on the chemical composition of the crystal. In most cases, the congruent composition is determined by growing an experimental series of crystals and determining the Curie temperature. The constancy of the Curie temperature along the entire length of the crystal boule indicates that the melt from which the crystal was grown is congruent. The reagents are weighed after setting the ratio of the components $\alpha Li_2O + (1 - \alpha)\ Nb_2O_5$, where α is the molar fraction of Li_2O, corresponding to the congruent composition. To ensure that the error in composition does not exceed ±0.025 mol.%, the absolute error per 100 g of component should not exceed: for Li_2CO_3 0.006 g, and for Nb_2O_5 0.023 g. This defines the requirements on the accuracy of weighing.

The weighed of reagents are then thoroughly mixed for several hours in a device of the 'drunk barrel' type. The mixture is then transferred into alundum vessels with a capacity approximately 1 dm^3, closed with alundum lids, and placed in the muffle furnace for annealing at 1000–1200°C for 8–10 h to ensure completion of the solid state reaction between the components of the compound and release of CO_2.

The crucible can be melted in a furnace with induction heating adding an extra charge several times to complete melting of the crucible. However, this method of filling the crucible with the melt leads to premature deformation of the crucible. To reduce crucible deformation, it is recommended to use a charge in the form of compacted tablets with the diameter equal approximately to the crucible diameter. The crucible can be melted in a single cycle in the muffle furnace. The crucible and the funnel above it are filled with the charge in such a manner that in melting the melt fully fills the crucible. The position of the crucible and the funnel is shown in Fig.72. The furnace is heated to approximately 1300°C, the charge melts and flows from the funnel into the crucible

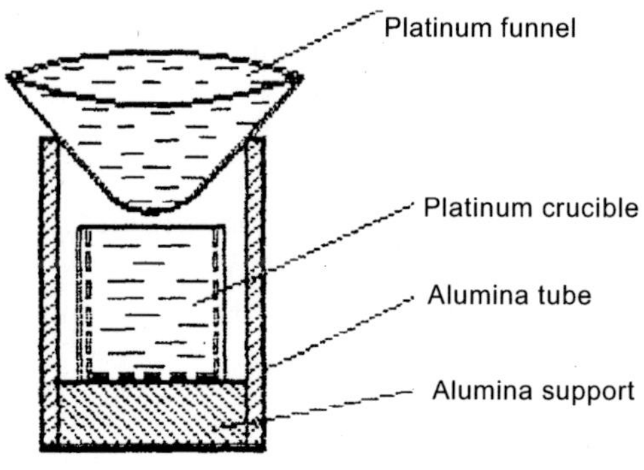

Fig.72 Diagram of filling a crucible with a melt in a muffle furnace using a platinum funnel.

and the melt is then held for several hours to homogenise and is then solidified.

The melt in the crucible should be colourless, transparent, and the polycrystalline edging representing possibly the $LiNb_3O_8$ phase should not be observed at the melt–crucible interface. The presence of this edging indicates disruption of the congruent nature of the melt. If these conditions are fulfilled, the crucible with the melt is ready for growing a crystal from it.

Positioning the crucible with the melt in the growth chamber. To grow crystals 50 mm in diameter, it is rational to use a platinum crucible 84 mm in diameter with a wall thickness of 2 mm. This crucible is placed in a compound alundum tube 100 mm in diameter with a wall thickness of 15–20 mm and positioned on a thick alundum base 40–60 mm thick. The lower part of the tube fully covers the crucible which is immersed by 1 cm from the upper edge of the tube. An inspection window in the form of a narrow slit 3 mm high and 20 mm wide is cut between the upper and lower parts of the tube. The inspection window is covered with transparent quartz class.

A platinum screen in the form of a cylinder 80 mm in diameter 100–120 mm long, with a wall thickness of 1 mm, is suspended on handles from the upper part of the tube. At the top, the cylinder is covered with a platinum disk with a hole for the holder of a seed crystal. 6–8 induction coil turns is used for the crucible, 3–4 turns for the screen. The withdrawal speed of the cylindrical part of the crystal decreases as the crystal is being pulled out at a rate from 4–6 mm/h to 2–3 mm/h in the lower part. The crystal is fully pulled into the screen without reducing the power emitted by the induction coil. Cooling from the solidification temperature to 1030–1050°C is carried out at a rate of 20–30 deg/h, further cooling is carried out at 100 deg/h at room temperature. It should be mentioned that the cooling conditions are selected by experiments taking into account the physico-chemical features of the crystal. The boule is separated from the seed crystal using tricot gloves and covered with cotton cloth to avoid local heating by the hands and formation of pyrocharges which could cause cracking of the boule.

24.2 Selection of crucible material

The platinum crucible is more suitable for growing lithium niobate crystals. Of all noble metals, platinum is least susceptible to interaction with the lithium niobate melt. The wall thickness of the crucible is 1.5–2 mm to reduce deformation during cooling of the melt. Doping platinum with small additions of zirconium or yttrium stabilises the block structure and approximately doubles the surface life of the crucible (Fraser, 1990). Usual additions of iridium or rhodium to platinum from which the crucible for growing lithium niobate is produced are not used because they cause colouring the crystal. This was observed for the first time by the author of this book.

24.3 Thermal screening of the crucible with the melt

Thermally insulated and reflecting materials, surrounding the crucible with the

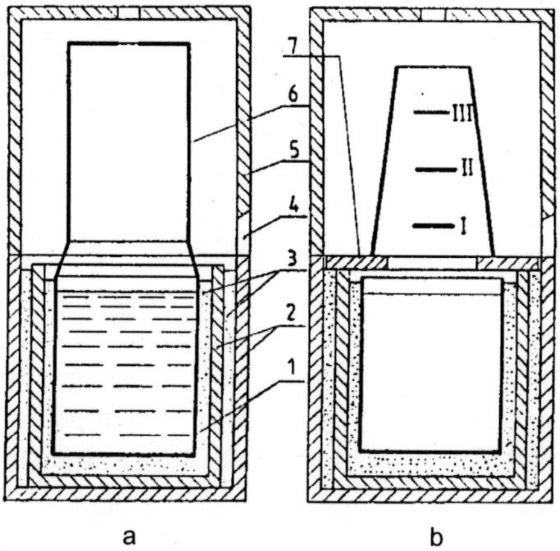

Fig.73 A schematic of the growth chamber: a) with a cylindrical screen; b) with a conical screen: 1) platinum crucible; 2) alumina (Al_2O_3) glass; 3) coarse-grain filler, Al_2O_3; 4) window for visual examination of crystal growth; 5) alumina lid; 6) platinum screen; 7) ceramic washer with a hole for the crystal holder. I, II,III are the position of the thermocouple in the crystal in measuring temperature gradients (Rubinina 1976).

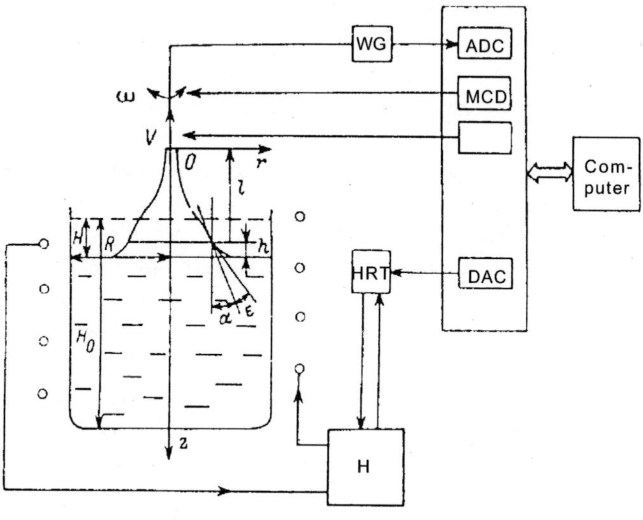

Fig.74 Diagram of inspection of the weight of a crystal during growth: WG) weight gauge of the crystal; DAC) digital-analogue converter; MCD) module for controlling the drive; ADC) analogue-digital converter; HRT) high-accuracy temperature regulator; H) source of heating power (Rossolenko et al, 1992).

melt and the growing crystal, should withstand the high temperature of more than 1300°C and should not interact with melt vapours. They should also ensure the required temperature distribution in the melt and above it. The typical systems are presented in Fig.73.

Air is an ideal atmosphere for growing lithium niobate. However, it should be noted that if there is a deficit of oxygen in the atmosphere niobium is partially reduced and the crystal becomes light brown in colour and its electrical conductivity increases. In the Czochralski method, the crucible with the melt is heated by resistance or rf induction heating, where the latter enables the most efficient control of crystal growth. Resistance furnaces are used to produce low temperature gradients in the growth zone which greatly reduce the probability of cracking of the crystal during growth. In systems with rf heating it is necessary to use thermal shields and additional heaters. Some of the types of these heaters are shown in Fig.73.

24.4 Control of diameter

Various methods are used for automatic control of the diameter of lithium niobate crystals during growth. Weight methods are used most widely. Sensors of crystal weight are most suitable when they are used in computer systems of controlling crystal growth. The block diagram in Fig.74 shows in the simplified form the structure of modern computer automatic control systems (ACS) used for growth of çrystals of various type from the melt. The digital code of the signal of the weight sensor is transferred to a computer where the effects of the heating power of the thermal zone, the drawing speed and the rotation rate of the crystal are calculated. After digital-analogue conversion the signals are transferred to the corresponding actuating devices (Rossolenko, et al., 1992).

24.5 Temperature gradients

Undoubtedly, the most important parameters in growing lithium niobate crystals by the Czochralski method are the axial and radial temperature gradients above and below the melt surface. A steep thermal gradient above the melt in the space into which the crystal is drawn may cause thermal stresses to form. The gradient below the melt surface should be such as to enable drawing a crystal at a high speed but without constitutional undercooling. However, convection takes place under these conditions and is accompanied by the formation of growth bands associated with the variation of the chemical composition of the solidified melt, if the congruent nature of the melt is disrupted.

The ideal form of the surface of the solidification front is the flat form which is determined by the ratio of axial and radial temperature gradients. If this ratio is high, the surface is flat. These problems were discussed in greater detail previously. The optimum parameters of crystal growth are usually selected by an empirical procedure. The gradient above the melt is regulated by modifying a surrounding thermal insulator or by introducing an additional heater. The gradient in the melt can be varied by regulating the vertical position of the crucible in the induction heater. Figure 75 shows typical temperature gradients in and above

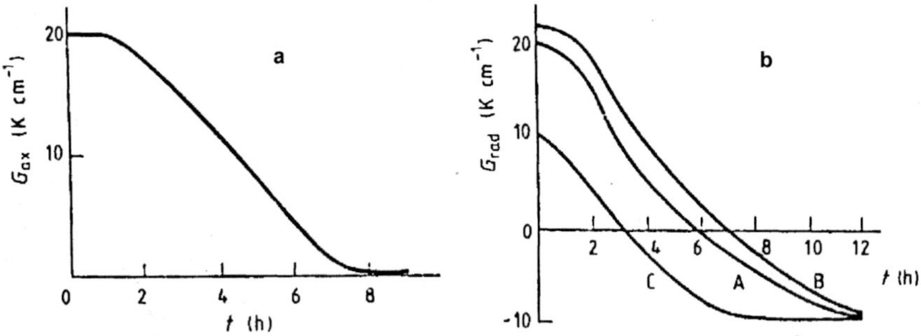

Fig.75 Change in the axial (a) and radial (b) temperature gradient in LN crystals being cooled: A, B, C corresponds to the positions of the crystal in the screen I, II, III shown in Fig.73.

the melt for thermal packing shown in Fig.73. According to the literature data, the axial temperature gradient above the melt in different growth systems vary from 20 to 100 deg/cm (Fraser, 1990).

24.6 Crystal growth rate

In the Czochralski method, the growth rate at which the crystal solidifies from the melt is usually a linear function of the axial gradient. In a general case, the growth rate is a sum of the drawing rates of the crystal and the rate of reduction of the level of the melt in the crucible. The growth rate (g) is linked with the drawing rate (p) by the following equation (Brandle, 1985):

$$g = p\left\{r / \left[r - (d/D)^2\right]\right\}, \tag{99}$$

where d an D are the diameters of the crystal and the crucible, respectively; r is the ratio of the density of the melt to crystal density. The literature often contains only the drawing rate and the reader is asked to draw a conclusion on the true rate from other values available. The examined rates are equal if the level of the melt in the crucible is regulated during crystal growth, as discussed below, or when the crystal diameter is more in comparison with the crucible diameter. In growing lithium niobate crystals, the ratio of the crystal and crucible diameters is usually 0.5–0.6 and the ratio of the densities approximately $r = 0.8$. Taking these values into account, the growth rate of the crystal is approximately 50% higher than the drawing rate. Fraser attempted to estimate the maximum possible growth rate of the crystal using the thermal balance equation (Runyan, 1965):

Latent heat in the interface + heat flux from the melt of the crystal = heat flux from the phase surface

$$L g \rho A + K_m (dT/dx_m) A = K_S (dT/dx_S) A, \tag{100}$$

where L is the latent heat of melting, T is temperature, g is the growth rate, ρ is crystal density, A is the cross section of the crystal, K_m and K_S is the thermal conductivity of the melt and the crystal, respectively, $dT/dx_{m,\,S}$ are the temperature gradients in the melt and the crystal, respectively.

This equation makes it possible estimate the maximum, physically attainable growth rate of the crystal and accelerator conditions under which the growth rate does not depend on the crystal diameter. The solution of the equation gives the maximum growth rate of lithium niobate of several cm/h. This value corresponds to the growth rates of certain elementary semiconductors, such as silicon or germanium and, theoretically, to purely congruently melting compounds. In practice, it is very difficult to achieve and maintain the precision congruent melt. One of the compound parts of the compound may be volatile and evaporate more rapidly from the melt thus disrupting congruency. In growing a crystal from a noncongruent melt, the constitutional undercooling sets the limit of the maximum growth rate. In practice, thermal insulation and the associated temperature gradient determine the maximum growth rate at which the cellular structure, associated with constitutional undercooling, does not form. An increase of the temperature gradient makes it possible to increase the growth rate but also intensifies thermal convection of the melt and, consequently, increases the intensity of the growth bands and sensitivity to cracking of the crystal as a result of thermal stresses. The drawing rates, published in the literature, are in most cases in the range 2–7 mm/h (Fraser). Usually, the growth rate of the crystal decreases with an increase of its diameter.

24.7 Rotation speed

The seed crystal in the Czochralski method must be rotated to maintain the thermal symmetry of the environment in crystal growth. At a high ratio of the crystal and crucible diameters the rotation of the crystal hydrodynamically induces a directional central flow from the bottom of the crucible to the growing crystal and, if this region has a high melt temperature, the crystal surface will be bent. Fraser (1990) compiled a table (Table 13) including the published technological parameters of growth of lithium niobate crystals: the growth and rotation

Table 13 Growth parameters of LN crystals in the Czochralski method (Fraser, 1990)

Pulling speed (mm/h)	Rotation speed (rpm)	Crystal diameter (mm)	Melt composition (mol% Li_2O)	Growth axes
2–3	8–15	100–125	48.6	*y, r*
5	20	50–75	48.6	*y, z, r*
3–4	–	75	48.6 (Fe)	*z*
3	6–30	83	48.6	*z, r*

Here *r* corresponds to [014] growth direction

rates, crystal diameters, composition and orientation. From this table, we present only the data for crystals of large diameter which are of special interest for industrial production.

25 Growth of Stoichiometric LN Crystals from a Crucible with Continuous Feed

It is well known that when growing a stoichiometric LN crystal, the composition of the melt in the crucible continuously changes and this prevents a crystal with homogeneous composition along the entire length of the boule from forming. Japanese scientists (Kitamura, et al., 1992, Kan, Sakamoto, et al., 1992) proposed a method of continuous feed of the crucible with the melt of the required composition. This method is used widely for growing crystals in semiconductor industry to increase the length of the crystal boule (Laudise, Parker, 1970). However, the technology of crystals of oxide compounds and, in particular, lithium niobate crystals, has its special features. As indicated by the Li_2O–Nb_2O_5 phase diagram (Fig.9), the stoichiometric LN crystal is in equilibrium with the melt enriched with lithium (58–60 mol.% Li_2O). Consequently, a stoichiometric LN crystal can be grown from a lithium-enriched melt.

The growth systems developed by Kitamaura and Kan were designed for maintaining a constant composition of the melt from which the stoichiometric LN crystal solidifies. The crystallisation system is shown schematically in Fig.76. A special feature of the growth system is the use of a two-section crucible consisting of internal and external crucibles connected in the lower part by a means of a thin tube which may be closed part of the start of crystal growth. The melt in the internal crucible is enriched with lithium from which a stoichiometric crystal is grown by the Czochralski method. The external crucible is filled with the stoichiometric melt. The stoichiometric LN powder is con-

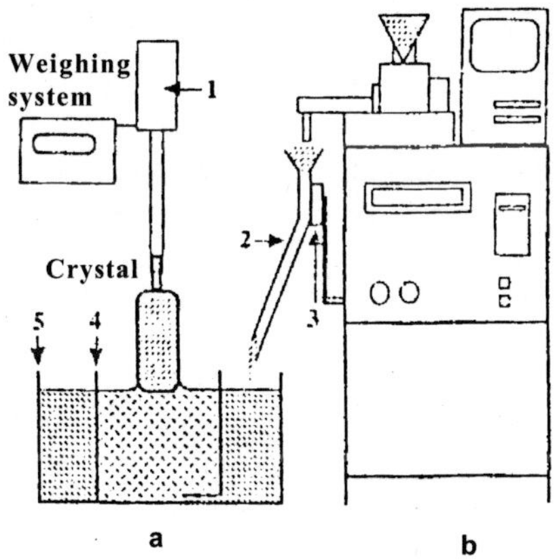

Fig.76 Diagram of a double crucible in the Czochralski method using an automatic system of supplying powder into the crucible: a) double crucible; b) automatic powder supply system. 1) loading cell; 2) Pt-tube; 3) piezovibrator; 4) inner crucible; 5) external crucible (Kitamura et al, 1992).

tinuously fed into the external crucible using an automatic powder feed system at the rate corresponding to the increase of the weight of the grown crystal.

The growth system consists of three parts: a double crucible, an automatic powder feed system, and Czochralski growth equipment. The double crucible consists of an external Pt-crucible with a diameter of 100 mm, 50 mm high and with a wall thickness of 1 mm. The internal Pt-crucible is 60 mm in diameter, 55 mm high, wall thickness 1 mm, and is placed in the centre of the external crucible. The authors paid special attention to the accuracy of automatic feed of the stoichiometric powder into the external crucible and to inspecting the weight of the growing crystal. Experimental results show that for charging it is most efficient to use LN powder with large particles produced by calcination at temperature slightly below the melting point. A special of growth equipment is the use of SiC heaters. Kan and Sakamoto (1992) used conventional high-frequency heating for this purpose. We shall discuss in greater detail crystal technology because it differs from conventional. Kitamaura, et al., paid special attention to the process of preparation of the powder for adding to the external crucible. The powder should be coarse-grained. The obtain this, the powder is compacted in a hydraulic press with a force of 1.5 t/cm^2. The compacted powder is heated to 1150°C to initiate grain growth, and the powder enriched with lithium is processed by the same procedure. The powder in the internal crucible is the first to melt. The level of the melt surface in the internal crucible remains constant to complete melting of the powder in the external crucible. Subsequently, the screen dividing the two crucibles is removed and the levels of the melt equalised. The seed crystals with zero orientation rotates at a speed of 5–10 rpm and is drawn at a speed of 1 mm/h. The diameter of the grown crystal is 37–40 mm. The coarse-grained powder of the stoichiometric LN is added to the external crucible at a rate of 6 g/h, as determined by monitoring the weight of the growing crystal. The grown crystals are annealed in an oxygen flow for several days at 1000°C. By measuring the parameter of the crystal lattice and the Curie temperature it was possible to obtain identical chemical composition along the entire length of the boule. This composition corresponded to the stoichiometric composition. Chemical etching showed that a large part of the crystal is of the single domain type, and the positive ends of the dipoles face the melt. Domains of opposite polarity were found only at the periphery (1–2 mm from the edge).

Thus, it may be concluded that, using the double crucible and the automatic powder feed system, it is possible to grow LN single crystals with a composition close to stoichiometric. The proposed method of growing single crystals has a number of advantages in comparison with conventional methods. Firstly, the level of the melt surface remains constant throughout the entire crystal growth process, i.e. the crystal growth at constant temperature gradients and constant convective flows of the melt in the crucible. The impurities are uniformly distributed throughout the entire volume of the crystal. In addition, large crystals can be grown from a small double crucible. In the examined case, the effec-

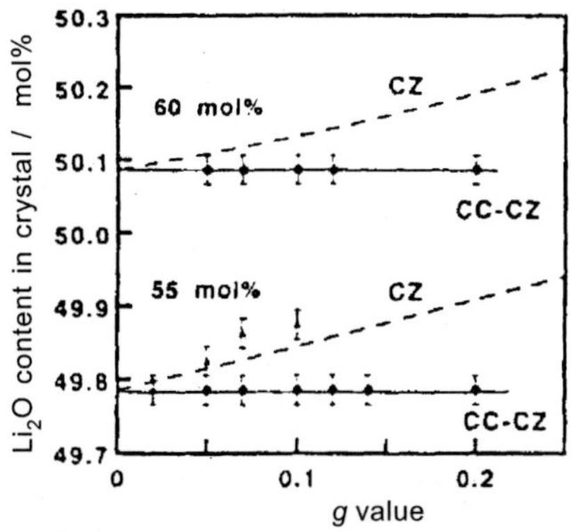

Fig.77 Change of the composition of the crystal with an increase of the melt *g* pulled from the crucible for crystals grown from compositions containing 55 and 60 mol.% Li_2O (Kitamura et al, 1992).

Table 14 Conditions of growth of stoichiometric LN crystals by the double crucible method

Drawing rate	1 mm/h
Rotation speed	10 rpm
Growth direction	<001>
Atmosphere	air
Initial composition of melt	48.6, 55.0, 60.0 mol.% Li_2O
Composition of material immersed in the external crucible	48.6, 49.8, 50.1 mol.% Li_2O

tive depth of the crucible was only 2.5 cm, i.e. the weight of the platinum crucible can be greatly reduced. Equipment constructed by Kan and Sakamoto slightly differs from that described previously. However, it is important to note the high stability of maintaining the thermal conditions of crystal growth and, consequently, the fact that the crystal diameter is constant. The above authors already achieved a constant chemical composition along the length of the boule, irrespective of the amount of the melt drawn from the crucible. This was obtained on the basis of the observed dependence of the parameters of the crystal lattice and the refractive index of the extraordinary beam from the melt composition. This is very important for crystal technology. Figure 77 shows the variation of composition along the length of the crystal boule in relation to the amount of drawn melt for the conventional Czochralski method and the Czochralski method using the double crucible.

Table 14 gives the technological parameters of the growing LN crystals by the double crucible method.

However, this crystal growth method has also serious shortcomings. They include the fact that technological equipment is greatly complicated by including a system for automatic feed of the initial powder into the external crucible with a high degree of accuracy equal to the weight of the drawn crystal. This circumstance is closely linked with the complicated process of preparing the initial powder of the coarse-grained fraction. The technological process is also complicated by long-term annealing the crystal at 1000°C in an oxygen flow. Attention should also be given to the low drawing rate of the crystal which is 1 mm/h. The quality of the crystal is influenced by the speed of flow of the melt through the platinum tube connecting both crucibles. In this respect, the double crucible method is analogue of growing crystals by the Stepanov method. Kan and Sakamoto confirm that only this method can be used to grow stoichiometric LN crystal homogeneous along the entire length of the boule.

26 Polarization of Lithium Niobate Crystals

The polarization process, i.e. ordering of electric dipoles formed by displaced niobium and lithium ions from symmetric crystallographic positions along the ferroelectric *C*-axis, is an important operation in the process of preparation of optoelectronic and acousto-optic devices using lithium niobate crystals. This operation is carried out at temperatures close to Curie temperature T_C and with the application of a constant electric field. Problems are caused by the fact that the T_C temperature of the crystals of congruent composition, equal to approximately 1170°C, is close to the melting point, T_m = 1260°C. The crystal is held under the field for one hour. The field is sustained during cooling the crystal below the Curie temperature and is removed prior to cooling the crystal at room temperature for 12–24 h. The cooling time is determined by the diameter and defective structure of the grown crystal.

To apply a field to the crystal, current-conducting electrodes are deposited at the outlet ends of the polar axis. The electrodes are usually in the form of platinum paste which is annealed into the crystal at 1200°C. The annealing-in time should be minimum to ensure burning out of the binder but it is important to avoid diffusion of platinum into the crystal. The electrodes can also be in the form of ceramic disks synthesized from LN powder or crystal sheets onto which platinum electrodes are attached to prevent melting of platinum with the crystal.

The strength of the electric field required for efficient polarization of the crystal at a temperature close to the Curie point is usually in the range 0.2–5 V/cm and in most cases is 1 V/cm (Fraser, 1990). To ensure complete polarization, the crystal should not contain any second phase inclusions and impurities which greatly increase conductivity. The level of current and the duration of application of the field must be controlled to avoid damage to the crystal as a result of diffusion of platinum, especially when a positive electrode is used.

Polarization is carried out in air or in an oxygen flow (approximately 2 litres/min). The diagram of crystal polarization equipment is shown in Fig.43. The monodomain state of the crystal can also be achieved by applying an electric field along the polar axis during growth. However, this method is suitable for crystals which are grown along the polar axis or close to it. In this case, the voltage of the applied field varies in the range 0.5–100 V to ensure passage of current through the insulating holder, the seed crystal, the growing crystal in the melt and also inversion of antiparallel domains. In this method, the passing current is maintained constant during crystal growth by increasing the applied field. The crystal grows in a stable manner when a negative voltage is applied to the crystal. Brice (1977) has reported that the formation of acicular domains takes place at secondary inclusions, and to produce a completely monodomain state it is necessary to grow a crystal free from inclusions.

To grow a monodomain crystal without applying the external electric field, molybdenum oxide (MoO_3) is added to the lithium niobate melt (Nassau, Levinstein, Loiacano, 1966). The mechanism of the effect of MoO_3 and the formation of the monodomain state has not been completely explained. The most probable mechanism is the formation, at the phase boundary, of a double electric layer which initiates ion ordering. The examined method does not make it possible to produce monodomain crystals and is used only for a limited range of growth conditions. In most cases when the crystal structure is disrupted by low-angle boundaries, the reversed local domains will be retained after polarization. These defects are smaller than 1 mm in length and several tens of microns wide and can exert a strong effect on the acoustic and electroacoustic properties. As already mentioned, the defects of this type are manifested in the form of ripple marks on the polished surface of the sheet produced for devices on the SAW.

When cooling the monodomain crystal below the Curie temperature a thin layer of negative domains forms on its positive end.

Evlanova (1978) established the following characteristic special features of the process of polarization of lithium niobate crystals. Nucleation of domains in the volume of the crystal takes place mainly on the (0001) surface in the form of wedges and in a smaller amount in the form of acicular domains. Domains on the $(000\overline{1})$ surface form at a later stage. The growth of domains in the direction of the applied field takes place a the rate considerably higher than the side movement of the walls. During repolarization, nuclei of new domains form and grow in the crystal volume. The domains grow both as a result of movement of the domain wall and by coalescence of individual domains. The process of repolarization ends at the $(000\overline{1})$ surface, and non-repolarised domains in the form of wedges remain in the vicinity of the surface. An increase of the strength of the field reduces the non-repolarised volume of the crystal but the crystal is not completely repolarised and an electric breakdown takes place.

Ivanova (1979) supplemented information on the formation of domains in lithium niobate crystals. In growth of the crystals along the polar axis domain formation is influenced only by the axial temperature gradient. For crystals with the growth direction normal to the polar *C*-axis a strong effect is exerted by the radial temperature gradient and, consequently, two main domains form in the symmetric temperature field. In growing the crystal along the direction forming an acute angle with the *C*-axis the formation of main domains depends on the ratio of the axial and radial gradients, Fig.78. The degree of unipolarity of the crystal increases with an increase of this ratio. The crystal grows monodomain if the projection of the axial gradient on the *C*-axis is larger than the projection of the radial gradient over the entire surface of the phase transition iso-

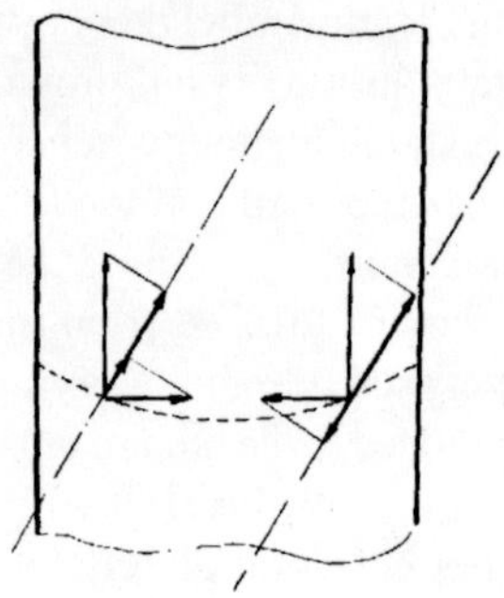

Fig.78 Diagram of formation of major domains in phase transition of a crystal grown under a oblique angle to the C axis: horizontal arrows indicate the direction G_{ird}, vertical – G_{axle}, dashed curve is the Curie isotherm, dot-and-dash curve is the C axis (Ivanova, 1979).

therm. The degree of unipolarity is increased by long-term holding of the crystal in the paraelectric phase and, consequently, by preventing the appearance of internal domains associated with the instability of the temperature field and with the presence of impurities.

The investigations carried out by Ivanova concerned with the effect of the growth conditions on the domain structure of the LN crystals show that to produce large monodomain elements from oblique sections of crystals grown from the congruent melt, it is necessary to ensure that the following conditions are satisfied (Ivanova, 1979):

1. A flat solidification front, i.e. a flat melting isotherm;
2. Maximum possible distance of the Curie isotherm from the phase boundary and its flat or slightly convex form. A flat Curie isotherm facilitates the process of formation of a monodomain crystal. It should be mentioned that these two requirements contain an internal contradiction from the viewpoint of their experimental realisation;
3. Constant thermal conditions throughout the entire growth process. These conditions are discussed in greater detail in another book by the author (Kuz'minov, 1987).

27. Formation of a Regular Domain Structure in Lithium Niobate Crystals

In the previous section, attention was given to polarization of LN crystals, i.e. transition of the LN crystal to the monodomain state. However, for a number of practical applications, which will be discussed later, it is necessary to use crystals with a regular domain structure (RDS). The RDS forms in the crystal by means of the periodic (with respect to the co-ordinate and time) external effect in the vicinity of the phase transition temperature. The interest in the RDS of ferroelectrics with a centrosymmetric paraphase which also include LN is the result of the possibility of controlling by alternating spatial modulation the physical properties of the crystal, i.e. paraelectric, piezoelectric and electro-optical properties, described by the tensors of an odd rank. These structures are characterised by the vector of spatial periodicity $K = \pi \mathrm{n}/d$, where n is the normal to the surface of the domain boundary; d is the domain size. The RDS can be produced during crystal growth as a result of formation of impurity growth bands during rotation and pulling of the crystal. The domain structure forms in cooling the crystal below the Curie temperature; the domain boundaries are growth bands and the distance between them is determined by the product of the pulling speed by the rotation period of the crystal. Antipov, et al. (Antipov, et al., 1985) proposed to form the RDS in a ready crystal by applying independent vector effects to it – an electric field and a temperature gradient. The orientation of the vector effects in relation to the spontaneous polarization axis of the crystal makes it possible to produce structures with the required vector K. The domain size given by the time periods of the variation of the sign of electric fields. To the form the RDS with the counterdirection P_S in adjacent domains at which $K \parallel P_S$, the electric field must be applied to the crystal parallel to the temperature gradient. In this case, the domain boundaries are perpendicular to the spontaneous polarization axis. Formation of the RDS with the counterdirection P_S in the LN crystals is possible due to the high electrical conductivity of the LN in the vicinity of the Curie temperature and formation of a screening charge at the domain boundaries. The RDS with $K \parallel P_S$ was formed by applying the electric field along the spontaneous polarization axis and normal to the temperature gradient. The following controlling technological parameters of the process were determined by experiments: the temperature gradient in the vicinity of the Curie point, current density and the strength of electric field at the domain boundary. The optimum parameters of RDS formation were specified and maintained by a computer so that it was possible to produce domains with the minimum thickness of the boundaries at a maximum value of K. However, in addition to the controlling effects, formation of the RDS is influenced strongly by the electrical conductivity of crystals and the concentration of point defects in the vicinity of the Curie temperature which determine the formation of the spatial charge of the screening field of spontaneous polarization. RDS formation is also influenced by the surface conductivity and uncompensated intracrystalline electric fields which distort the form of the domain structure.

Domain structures were developed by selective etching the polar faces of the crystal and examined under a microscope. Examination confirmed the structure of counterdomains whose size is determined by the switching time of the external field. It should be mentioned that in switching the polarising field the spontaneous polarization direction changes not only along the Z axis but also along the Y axis. This was observed on the YZ surface of the crystal. The latter circumstance is the consequence of special features of the crystal structure of the LN.

Examination under a microscope of optically indiscernible 180° domains in the LN crystal is possible as a result of a phase contrast formed in the region of the domain boundary due to internal electric fields. The thickness of the domain boundaries is determined by the gradient of the refractive index in the transition region. The effective thickness of the domain boundary in the LN crystal is 40–60 μm for $K \parallel P_S$ and 2–5 μm for $K \perp P_S$ (the results obtained by Antipov et al., 1985). As an example, Fig.79 shows a photograph of the periodic domain structure of an LN with domains 20 μm in size.

The resistance of the RDS to external effects was examined by determining the variation of the diffraction pattern from the phase grating formed in the RDS crystal, as a result of the cyclic effect of an alternating electric field. The variation of the intensity of the diffracted light observed when 10^6 pulses of a field with a strength of $5 \cdot 10^5$ V m^{-1} were applied to the crystal did not exceed 1%.

Lithium niobate crystals with the RDS can be used for light modulation and conversion (Blistanov, et al., 1986). When an external electric field is applied to a crystal with the RDS, the phase diffraction grating with the period equal to the double domain thickness will form in it as a result of differences in the sign of the electro-optic coefficient r_{33} in adjacent domains. The intensity of light in the diffraction maximum in relation to the control voltage is shown in Fig.80. The maximum intensity at zero maximum, equal to 98%, was obtained at a voltage of 5.3 kV. The use of LN crystals with the RDS for modulating radiation makes it possible to ensure highly efficient control of the light intensity reaching 100 MHz. For the sake of objectivity, it should be said that the control voltage is considerably higher, and the operating speed is too low in comparison with conventional electro-optic modulators.

The quasi-synchronous process of second harmonics generation can be realised for ferroelectrics with RDS. This phenomenon was examined in greater detail by the authors (Prokhorov, Kuz'minov, 1990).

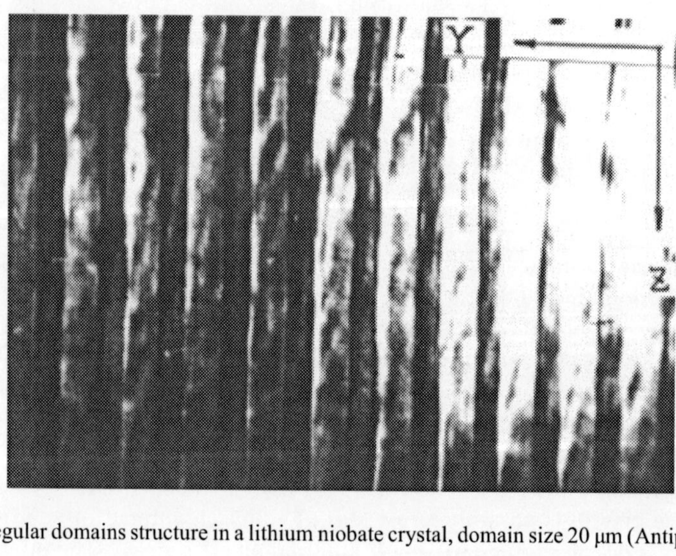

Fig.79 Regular domains structure in a lithium niobate crystal, domain size 20 μm (Antipov et al, 1985).

28. Determination of the Suitability of Crystals for Producing Devices Based on Them

The term 'suitability' is broader than the term 'quality'. If special attention in scientific studies is given to the physical properties of matter, in industrial production it is necessary to examine the relationship between the physical parameters of the matter and the economics of its production. We shall use the term 'quality' to describe the physical parameters of the crystal in relationship with the properties of devices produced using these crystals, and the term 'suitability' will be used to describe the physical properties and economic parameters of crystal growth for construction of these devices.

The term 'quality' is used for subjective evaluation of the crystal. The concepts 'good or poor' quality are used in most cases. This broad concept is not suitable for characterising the crystal for technical application. Therefore, the strict definition of quality such as the "identity of the properties of the component with respect to the requirements imposed on it" has been proposed (Fenske, 1988). As a result of the purposefulness of using the 'quality' concept, it relates to the correspondence of the set of the relevant properties of the given object for the given application. This definition may result in a paradox, i.e. an absolutely defect-free crystal can be classified as being of low quality, if a device can be produced, without losing application, on the basis of crystals with a specific concentration of structural defects.

Examining the relationship between the specific properties of the crystal and the required economic factors, for example, the low production cost of a specific component made of LN, it is possible to determine the requirements of the quality of the crystal in any stage of the technological process. As the accuracy for the which the requirements of quality are determined in each stage of technological process increases the efficiency of controlling the produc-

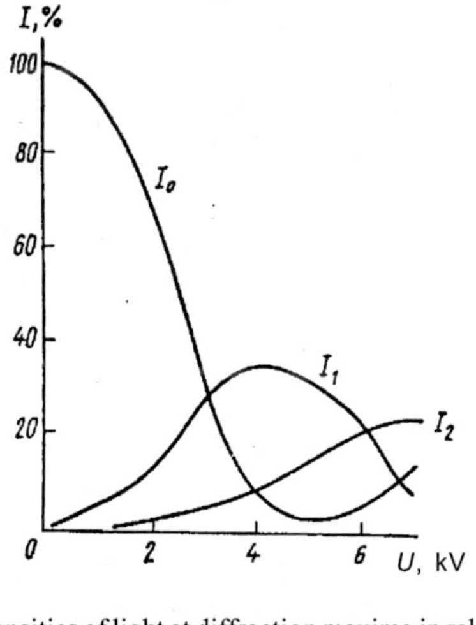

Fig.80 I_0, I_1, I_2 are the intensities of light at diffraction maxima in relation to the control voltage on the lithium niobate crystal with RDS (Blistanov et al 1986).

tion cost of components by using scientifically justified measures also increases. The quality control can be efficient if it combines organisational measures with extensive scientific analysis of the physical properties of the crystal and economic parameters of its production.

In production, it is often not possible to determine the properties of the crystal to the required extent. This circumstance leads to estimates exceeding the requirements thus increasing the production costs. Therefore, one of the main directions of scientific investigations aimed at reducing the production cost is to design specialised devices and measuring systems for determining the quality of the crystal and also corresponding measurement methods.

The second scientific direction includes examination of the technological process and continuous analysis of information. This makes the process science-intensive and determines the directions and measures of correcting the technology to obtain products of the required quality.

The third scientific direction is to find new procedures and methods of examination which simplify technology and reduce expenditure. The introduction of new elements into the current technological process requires matching all links from the investigator to the operator of growth mounting, including services of specialised analysis and completed components.

The high efficiency of scientific investigations in this direction can be ensured only by efficient matching of all effects under a single principle whose role may be played by a department dealing with quality problems. A system of this type has been used successfully in producing lithium niobate crystals for frequency selectors of television sets. This system controls all technological processes, from analysis of initial chemical reagents and crystal growth to the production of piezosheets.

As an example, we shall describe an integral system for ensuring the quality of LN crystals for frequency selectors of TV sets, Fig. 81.

28.1 Control of the process of production of lithium niobate crystals for TV filters on surface acoustic waves

Figure 81 shows a scientifically justified system of ensuring the quality of LN crystals in all technological stages of its production. The schema shows all flows of matter and information accompanying a batch of initial reagents, during the crystal production process. Special attention should be given to the fact that the decision is taken in eight stages on the feasibility of further processing of a given batch of initial reagents. If positive results are obtained in testing on the basis of the established standards, a batch of reagents travels further along the technological chain. If the requirements are not fulfilled, decisions are taken on the feasibility of further use of the initial material, or returning the crystal to remelting, carrying out further investigations and also on possible consequences in further processing of the material, and the rest of the technological process to complete investigations of consequences of the observed disruption. The criteria of efficiency of the work of the department dealing with quality is the yield (in per cent) of suitable devices with acceptable losses.

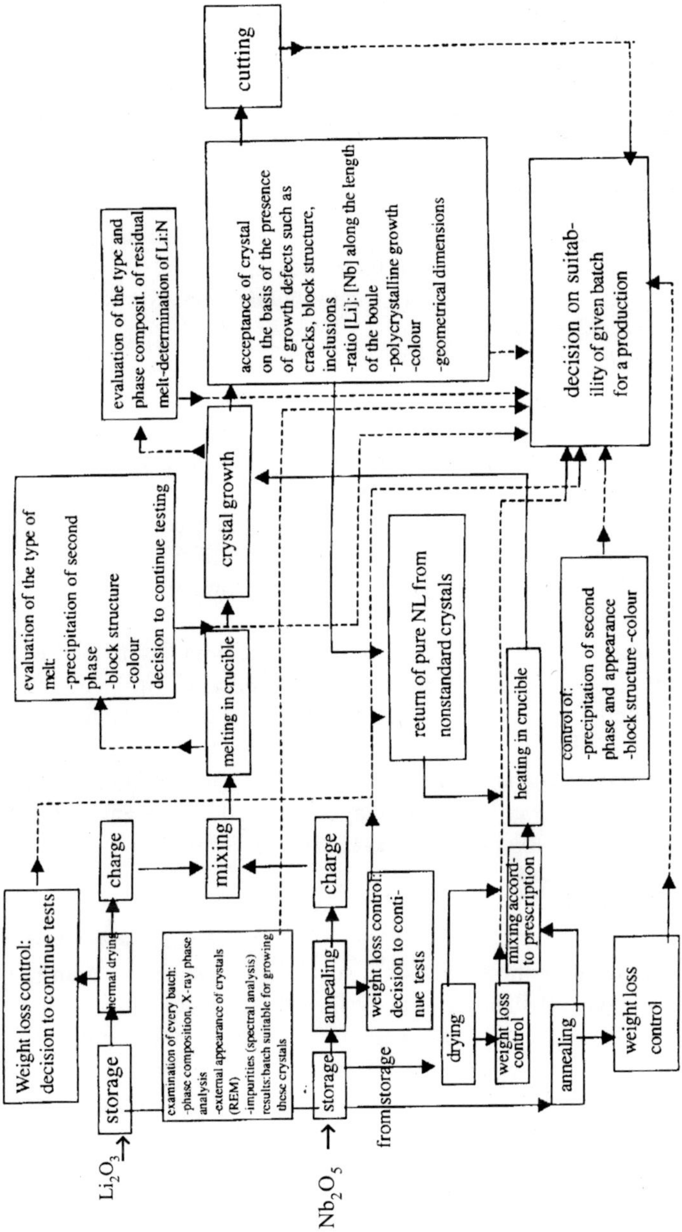

Fig. 81 Quality stabilization system (solid lines matter flow, broken lines information flow).

28.2 Requirements on LN crystal sheets for TB field tests

LN sheets suitable for producing filters for the intermediate frequency channel of TV sets should satisfy physical, geometrical, technological and economic requirements.

This type of filter (MSF 38.9) is produced using the crystallographic cut $(01\bar{4})$ or in other notations 38° ZXL or 128° XYL. This cut has a number of advantages in relation to the previously used Y-cut. In this cut, the degree of suppression of volume waves is considerably higher so that the filter area can be halved. The possibility of growing the crystal in this direction was established for the first in 1971 (Byer, et al., 1974).

The LN sheet is characterised by the following properties typical of the given cut:

– surface acoustic waves (SAW) propagate along the X axis
– the speed of sound of the SAW is 3995 m/s
– the electromechanical coupling factor is 5.5%.

In addition to the values given by the cut, the sheet should be characterised by low attenuation of acoustic waves and low dielectric losses.

The selected sound guide materials has the required properties if the technological process of production of piezosheets causes no disruptions in the crystal structure and additional essential conditions in fabrication of the device are fulfilled.

Slobodnik (1981) established that the gas medium has no marked effect on the acoustic losses in the frequency range of operation of the television filter. The calculated losses are equal to 0.02 dB which is two orders of magnitude lower than the allowance for the amplitude–frequency characteristic (AFC).

Technical documentation for the standard MSF-38.9 filter specifies the allowance for the frequency of the image carrier at a frequency of 38 900 kHz of the AFC filter as lower than 50 kHz. The equation

$$\Delta V = V_0 \, (\Delta f/f) \qquad (101)$$

is used to determine the permissible scatter of the speed of sound of $\pm$ 6 m/s at a mean speed of 3970 m/s. This allowance also includes errors of phototemplate and the variation of the speed of sound as a result of inaccuracy of the crystallographic orientation of the sheet. Therefore, the allowance for the speed of sound on the three surfaces is assumed to be equal to $\pm$ 2 m/s.

In this frequency range with a mean length of the acoustic wave of 100 μm it is not necessary to polish the surface of the sound guide because the surface roughness of the ground surface is in the range $R_a < 5$ μm which is considerably lower than the length of the acoustic wave. Attenuation is described by the expression (Grabe, et al., 1939):

$$\Gamma \sim \omega^5 a^2 \qquad (102)$$

Extrapolation of the experimental results, obtained by Slobodnik, to the frequency range 40 MHz leads to negligible losses, lower than the sensitivity threshold of the loss measuring devices, 0.01–0.1 dB.

The scattering centres in the active volume of the device based on SAW convert the energy of surface waves to the energy of volume waves. Their speeds differ and this distorts the amplitude- and phase–frequency characteristics. The volume scattering centres in lithium niobate crystals can be inclusions of the second phase or domain boundaries. Equation (102) shows that single inclusions with a length of up to several microns have no effect on the acoustic properties of the sound guide.

However, it should be mentioned that the speed of the sound wave is strongly influenced by the variation of the density along the crystal boule which occurs when growing the crystal from a non-congruent melt. This situation can be described easily in the following manner. The phase speed of the transverse elastic waves in a homogeneous isotropic solid medium is determined by the expression

$$V = (G/\rho)^{1/2} \tag{103}$$

where G is the shear modulus of the medium, ρ is its density. It is assumed that the composition of the crystal along the boule changes from congruent to stoichiometric. In practice, the composition of the crystal boule can vary over a wide range. This occurs especially in the multiple growth of crystals from large crucibles which carry several kilograms of the melt. This practice is used in a number of Russian plants. Simple transformation of the equation (103) makes it possible to evaluate the difference of the speeds for the congruent and stoichiometric crystals

$$V_c - V_s = V_c\,[1 - (\rho_c/\rho_s)^{1/2}] \tag{104}$$

and substituting the numerical values into equation (104), we determine the difference of the speeds: $\rho_c = 4.646$ g/cm^3, $\rho_s = 4.634$ g/cm^3, $V = 3995$ m/s, $\Delta V = 5$ m/s. In practice, the scatter of the speed of SAWs is several times greater than the determined value and, therefore, the constancy of composition along the length of the crystal boule is of considerable importance. It should be mentioned that the fluctuations of chemical composition in the range of the growth boule may be considerable and also lead to the scatter of the speeds of the SAWs.

In producing devices from a crystal sheet, it is important to fulfil the requirements resulting from apparatus and physico-chemical allowances for devices. The first group of requirements relates to the geometrical parameters of the sheet. These requirements are summarised in Table 15 and do not require additional explanation, with the exception of the surface finish. Taking into account the attenuation of the SAWs, it will be sufficient to confine requirements to the ground surface, but the photolithographic process and also contacting cannot be carried out on such a surface. At a thickness of the layer of the aluminium structure of 200–300 nm any depression of this size leads

Table 15 Technical requirements on lithium niobate substrates for television filters

Top side	polished
Reverse side	ground
Diameter	51 1 mm
Thickness	500 ± –25 μm
Deviation from parallelity	< 25 μm
Bending of sheets	< 45 μm
Orientation of the normal to the working surface	$[0\bar{1}4]$ 38° ZXL
Error in the direction of the normal	< ± 30'
Facet-direction	⊥ (X), ± 1°
Length	20-25 mm
Edge defects	< 1 mm
Cracks	no
Roughness - top side	< 30 nm
Reverse side	< 2 ± 0.5 μm
Scratches - top side - right hand side	none
Reverse side	a total of < 100 mm
Contamination	no insoluble contaminants

to rupture and disruption of the interdigital strip electrodes converter.

28.3 Post growth control of crystal boules

After growing the crystal and forming a monodomain structure (polarization), the boule is inspected prior to machining to determine several parameters:

– geometrical dimensions;
– nature of defects;
– resistance to mechanical effects.

In addition to these operations, it is also necessary to evaluate the residual melt. This must be carried out because of two reasons. Firstly, the nonstandard melt must not be returned to the process. Secondly, the residual melt shows earlier (than the crystal) disruptions of stoichiometry and purity of the initial substance.

In addition to measurements of the geometrical dimensions, the first stage also consists of detection of cracks and cleaves. As shown in Table 15, they most not form on the working surface of the crystal. Attention must be given to the appearance and size of the facets. The size of the facet is a measure of fulfilling the temperature conditions inside the thermal growth zone. Excessive expansion of the facets at the end of the crystal indicates a large reduction of the temperature gradients at the growth run of the crystal and restricts the useful volume of the crystal. This part of the crystal can be sent for remelting. The crystal boule should have a smooth, clean and transparent surface.

In the second stage, defects in the crystal bulk are detected. The crystal is examined in a strong parallel light beam. Tilting the light source and examining the crystal in the perpendicular direction shows optical heterogeneities, as already mentioned. This method can be used to found the following defects:

a) grain boundaries and small-angle grain boundaries;

b) inclusions of the second phase which can be lithium triniobate;
c) polycrystalline regions;
d) subsurface cracks;
e) polydomain regions.

Type *a* defects are found in the form of paired lines on the crystal surface. The volume of the crystal between the defects from the remaining volume in the direction of the crystallographic axes which appear on x-ray topograms in the form of dark and light regions with a difference of two angular minutes between the axes.

Type *b* inclusions are visible as flickering points in the crystal volume. Larger disruptions of technology or contamination of the initial reagents are visible as turpid regions inside the crystal.

Polycrystalline regions, defects of type *c*, usually fracture during cooling of the crystal. Subsurface cracks (defects *d*) form if the thermal conditions in the upper part of the growth chambers are not adhered to. In the polydomain regions of the crystal the light scatters on the domain walls. The polydomain regions form if current is interrupted during the formation of a monodomain structure and are not associated with the real structure of the crystal.

The third stage is splitting of the crystal into parts to remove nonstandard regions. This operation is usually carried out in a sawing machine with an external cutting edge. The crystals which do not crack in this operation have a low level of residual internal stresses and are referred to as 'cold'. All subsequent operations of cutting and grinding the external surface of the boule are less critical and usually do not cause additional cracking.

29. A Method of Reducing Deformation of the Crucible during Crystal Growth by the Czochralski Method

The crystals of oxide compounds are grown in platinum or iridium crucibles. The design of crucibles is simplified using also alloys of noble metals, for example, a platinum–rhodium alloy. However, when growing the lithium niobate or tantalate from platinum–rhodium crucibles, the crystals assumed intense brown colour as a result of implantation of rhodium into the crystal lattice. The increase of the strength of the crucibles by alloying, diffusion bonding or by increasing the wall thickness does not lead to desirable results.

Deformation is especially intensive in rapid cooling the crucible with the melt. The need to interrupt the process in the initial stage of crystal growth arises in, for example, inefficient seeding or as a result of malfunction of the lifting or rotation mechanisms. The difference of the coefficients of thermal expansion of the crucible material and the molten substance results in the formation of mechanical stresses in the crucible. When reaching the yield limit of the material, the crucible is irreversibly deformed. The most frequent type of deformation is the expansion of the lower part of the crucible. The cylindrical form of the crucible can be partially restored by rolling or pressing in. However, these methods shorten the service life of the crucibles and are associated with the loss of expensive metals.

It should also be noted that when using deformed crucibles, the crystal growth process cannot be reproduced: the symmetry of the thermal field is disrupted, the technological conditions of crystal growth change. Deformation can be greatly reduced by pulling the entire melt from the crucible. It has already been mentioned that when using a congruent composition, determined with a high degree of accuracy, up to 98% of the melt can be pulled out.

A method of cooling thin-walled crucibles filled with the melt is described below. This method makes it possible to avoid deformation of the crucible (Fenske, 1988). The proposed method is suitable for materials whose density greatly increases during solidification, for example, for lithium niobate crystals. The method proposed below increases the service life of the crucibles and shortens the cooling time of the melt.

The physical principle of deformation of the crucible is that the solidified melt is tightly compressed to the crucible walls and a cavity forms in the centre of the crucible. The thick layer of the solidified melt is a strong stop for the crucible wall. During cooling the area of the shell of the crucible decreases, the cylindrical crucible is tightly pressed to the ridge wall made of the polycrystalline material. Tensile stresses for irrespective of the cooling rate and their value exceeds the limiting plastic tension of platinum and the cold crucible is stretched. To prevent the tensile stresses acting on the crucible, it is necessary to form a slit between the solidified melt and the crucible walls. This slit can be empty or filled with the porous material of the melt which has no appreciable resistance in compression of the crucible walls during solidification.

By applying technological operations in a specific sequence, it is possible to produce such a space between the crucible wall and the solidified melt. This sequence of the operations can be described as follows. The crucible with the melt solidify at the normal rate of temperature decrease to complete solidification of the melt. The temperature of the crucible walls should then be rapidly increased. At temperatures considerably higher than the melting point the lithium niobate melts in the vicinity of the crucible walls. The volume of the resultant melt is larger than the slit around the internal surface of the crucible wall and the melt fills the depression in the central part of the crucible in which the crystal block has the temperature lower than the melting point. The melt fills the cavity in the centre of the crucible and solidifies. When this process is completed, the temperature of the crucible wall must be rapidly reduced. The remaining melt solidifies during this process. Since part of the melt left the space around the wall, a porous structure of polycrystalline lithium niobate forms in this space. This intermediate layer has no influence on the compression of the crucible wall and, consequently, protects the crucible against deformation. We shall describe a typical example for lithium niobate. A crucible with a wall 2 mm thick and 90 mm in diameter with a lithium niobate melt expands from 0.3 to 0.6 mm during one hour of operation during solidification at a rate of 100 K/h. In the proposed method, the melt temperature is reduced at a rate of 60 K/min to 1230°C and the melt is then held at this temperature for one hour.

The melt with a density of 3.57 g/cm^3 solidifies (the density of the solid phase is 4.5 g/cm^3). A depression forms in the centre of the crystal block. The rapid increase of the temperature of the crucible wall at a rate higher than 500 K/min to temperatures 20 K above the melting point leads to filling of the depression by the melt over a period of approximately 5 min. When the depression is filled with the melt the temperature rapidly decreases by 50 K. The intermediate layer formed in this manner with a thickness of approximate 5 mm is partially free from the material and partially filled with the porous material with a large fraction of air holes. After these operations, the crucible is cooled at a rate of 200 K/h without residual deformation.

Conclusions

It is obvious that this book does not deal with all aspects of the technology of lithium niobate crystals. The wish to grow this unique crystal of high optical quality and with the required physical properties forces technologists and physicists to conduct investigations into the technology and effect of various physical and chemical factors on the properties of this crystal. Obviously, lucky discoveries and inventions will be made and used in practice. It should be noted that detailed examination of the technology of lithium niobate crystals enriches the technology of other crystals and is of general nature. The current technological level corresponds to the requirements on the industrial application of this crystal but this does not indicate the end of further progress in this area.

REFERENCES

Abrahams S.C., 1990, in: Properties of Lithium Niobate, EMIS, Datareviews, ser, No.5, 3.

Abrahams S.C. and Marsh P., 1988, Acta Cryst., B, 42, 61.

Abrahams S.C., Reddy J.M., Bernstein J.L., 1966, J. Phys. Chem. Solids, 27, 997.

Arminise M.N., Canali C., De Sario M. et al., 1983, J. Appl. Phys., 54, 6223.

Ballman A.A., Levinstein H.J., Capio C.D., Brown H., 1967, J. Am. Ceramic. Soc., 50, 657.

Bollmann W., Schlothauer K., Zogal J., 1977, Kristall und Technik, 11, 2, 1327-1332.

Bollmann W. 1977, Phys. Stat. Solid. (a), 40, 83-91.

Bollmann W. and Stohr H.J., 1977, Phys. Stat. Solid. (a), 39, 477.

Bol'shakov S.A., Klyuev V.P., Lyapushkin N.N. and Lyubimov A.P., 1969, Inorganic Materials, 5, 5, 969

Bordui P.F., Norwood R.G., Bird C.D., Calver G.D. 1991, J. Cryst. Growth, 118, 61-68.

Born E., Willbald E., Veith R., 1982, Ultrasonics Symposium, 367.

Brandle C.D., 1985, in: Crystal Growth of Electronics Materials, North Holland, Amsterdam.

Brice J.C., 1968, J. Cryst. Growth 2, 395.

Brice J.C., 1977, J. Cryst. Growth 42, 427-30.

Bridenbaugh P.M., 1971, J. Appl. Phys. 42, 427-30.

Bridenbaugh P.M., 1973, J. Cryst Growth., 19, 45-52.

Buraches S.F., Timan B.L., Bondar' V.G., Dubovik M.F., 1978, Production and examination of single crystals, Scientific Research Institute of Single Crystals, Khar'kov, No.1, p.6

Buraches S.F., 1981, Examination and automation of the process of growth of lithium niobate crystals, Kha'rkov, Ukraine.

Burton J.A., Prim R.C., Slichter W.P., 1953, J. Chem.Phys. 21, 1987.

Byer R.L., Herbst R.L., Feigelson R.S., Kway W.L., 1974, Opt. Commun., 12, 427.

Carruthers J. R., 1976, J. Cryst. Growth 36, 212.

Carruthers J. R., Peterson G.E., Bridenbaugh P.M., Grasso M., 1971, J. Appl. Phys., 42, 1846-51.

Chernov A.A., 1980, Modern Crystallography, vol.3, Nauka, Moscow.

Dhar A. and Mansingh A., 1990, J. Appl. Phys., 68, 5604.

Donnerberg H., Tomlinson S.M., Catlow C.R.F., Schirmer O.F., 1989, Phys. Rev. B, 40, 1909.

D'yakov V.A., Shumov D.P., Rashkovich L.N., Aleksandrovskii A.L., 1985, Fiz. Tverd Tela, 49, 2418.

Evlanova N.F., 1978, Domain Structure of Single crystals of Lithium Metaniobate Grown by the Czochralski Method, Dissertation, Moscow State University, Moscow.

Fairbrother F., 1972, The Chemistry of Niobium and Tantalum, Russian translation, Khimiya, Moscow.

Fairbrother F., 1967, The Chemistry of Niobium and Tantalum, Elsevier Publishing Company, Amsterdam/London/New York.

Fay H., Alford J., Dess H.M., 1968, Appl. Phys. Letters, 12, 89-92.

Fedulov S.A., Shapiro Z.I., Ladyzhinskii P.B., 1965, Kristallografiya, 10, 268.

Fenske M., 1988, Development of an Automated Process of Growing Lithium Niobate Crystals under Industrial Production Conditions, dissertation, General Physics Institute, Academy of Sciences of the USSR, Moscow.

Fenske M. and Kuz'minov Yu.S., 1988, Determination of the Chemical Composition of

Lithium Niobate Crystals from the Curie Temperature, Preprint No.48, General Physics Institute, Russian Academy of Sciences, Moscow.

Feng Xi-Qi, Ying Ji-Feng, Wang Jin-Chang, Liu Jian-Cheng, 1988, Acta Physics Sin. (China), 37, 2062-7.

Fraser M. 1990, in: Properties of Lithium Niobate, EMIS Datareviews Series 5, 207.

Fraser M., 1990, in: Properties of Lithium Niobate, EMIS Datareviews Series 5, 220.

Furukawa Y., Sato M., Ninanda F., Inj K., 1990, J. Cryst. Growth, 99, 832.

Gallagher P.K., O'Bryan H.M., 1988, J. Am. Ceram. Soc., 71(1), C56-59.

Garmash V.M., Zinkina G.P. and Lazareva V.B., 1972, in: Crystal Growth, 9, Nauka, Moscow.

Grabmaier B.C., Otto F., 1986, J. Cryst. Growth 79, 682.

Grabmaier B.C., Otto F., 1986, SPIE Proc. 651, 2.

Grabmaier B.C., Wersing W., Koestler W., 1990, J. Cryst. Growth 110, 339-47.

Grab G., Kubaschewski O. et al. 1939, Z. Elektrochemie 45, 855.

Guenais B., Baudet M., Minier M., Le Cun M., 1981, Mat. Res. Bull.643-53.

Haycock P.W. Towned P.D., 1986, Appl. Phys. Letters, 48, 698-70.

Hennigsen T., Singh N.B. 1989, J. Cryst. Growth, 96, 114-8.

Herrington J.R., Dischler B., Rauber A., Schneider J., 1973, Solid State Commun., 12, 351.

Hinkov V., 1990, Properties of Lithium Niobate, EMIS Datareviews Series, No.5, An Inspec Publication, 94.

Hurle D.T.J., 1961, Solid State Electron., 3, 37.

Hu L.J. and Chang Y.H., 1991, J. Cryst. Growth., 114, 191-197.

Hu L.J., Chang Y.H., Yen F.S., 1991, J. Appl. Phys., 69, (11), 7635.

Idenbom V.M. and Tomilovskii G.E., 1958, Kristallografiya, 3, 593.

Iliev K., Nikolov V., Peshev P., Piekarezyk W., 1991 J. Cryst. Growth 108, 219-224.

Ivanova Z.I., 1979, Examination of the Parametric Generation and Amplification in Lithium Niobate Crystals of High Optical Quality, dissertation, Moscow State University, Moscow.

Jin. W. and Tsukamoto K., 1992, J. Cryst. Growth., 123, 327-332.

Jin. W., Jin J. Zhoh W., 1987, Chinese Phys. Letters, 4, 181.

Jorgenson P.J. and Bartlett R.W., 1969, J. Phys. Chem. Solid., 30, 2639.

Kan S., Sakamoto M., Okano Y., Hoshikawa K., Fukada T., 1992, J. Cryst. Growth, 119, 215-20.

Kitamura K., Yamamoto J.K., Iyi N., Kimura S., Hayashi T., 1992, J. Cryst. Growth, 116, 327-32.

Kobayashi N., 1981, J. Cryst. Growth 55, 339.

Kofstad P., 1972, Nonstoichiometry, Diffusion and Electrical Conductivity in Binary Metal Oxides, Russian translation, Mir, Moscow.

Kofstad P., 1972, Nonstoichiometry, Diffusion and Electrical Conductivity in Binary Metal Oxides, Wiley-Interscience, New York.

Kovacs L. and Polgar K., 1986, Cryst. Res. Technol., 42 K, 101-104.

Kovacs L. and Foldvari I., 1990, Properties of Lithium Niobate, EMIS Dataviews Series, N 5, 189.

Kumada N., Ozawa N., Muto F., Kibjura N., 1985, J. Solid State Chem., 57, 267.

Kushibiki J., Takahashi H., Kabayashi T., Chubachi N., 1991, Appl. Phys. Letters, 58, 2622-24.

Kuz'minov Yu.S., 1975, Lithium Niobate and Tantalate – Materials for Nonlinear Optics, Nauka, Moscow.

Kuz'minov Yu.S., 1987, Electrooptical and Nonlinear Optical Lithium Niobate Crystal, Nauka, Moscow.

Kuz'minov Y.S. and Osiko V.V., 1991, Ferroelectrics, 112, 105-113.

Laudise A. and Parker R.L., 1970, The Growth of Single Crystals. Solid State Physics, vol. 25, Academic Press., New York and London.

Lodiz Z. and Parker R., 1974, Crystal Growth, Russian translation, Mir, Moscow.
Lerner P., Legras C., Dumas J.P., 1968, J. Cryst. Growth, 3/4, 231.
Lee S.H., Kim Y.J., Cho S.H., Yoo E.P., 1992, J. Cryst. Growth., 125, 175-180.
Lewis M.F., West C.L., 1990, in: Properties of Lithium Niobate, EMIS Datareviews Series, 5, 104-105
Lines M.E. and Glass A.M., 1979, Principles and Applications of Ferroelectrics and Related Materials, Clarendon, Oxford.
Lundberg M., 1971, Acta Chemica Scandinavica, 25, 3337.
Melankholin N.M., 1970, Methods of Examining the Optical Properties of Crystals, Nauka, Moscow.
Mileshina V.A., Belugina N.V., Rakova E.B., Nagdasarov Kh.S., Zakharov N.D., Rozhanskii V.N., 1984, Kristallografiya, 29, 1163.
Nassau K., Levinstein H.J., Loiacano G.M., 1966, J. Phys. Chem. Solids, 27, 983.
Nassau K., Levinstein H.J., Loiacano G.M., 1966, J. Phys. Chem. Solids, 27, 989.
Nikolov V., Iliev K., Peshev P., 1988, J. Cryst. Growth 89, 313.
Nikolov V. Iliev K., Peshev P., 1988, J. Cryst. Growth 89, 324.
Niizeki N., Yamada T., Toyoda H., 1967, Jap. J. Appl. Phys., 6, 318.
Non-Stoichiometric Compounds, Edited by L. Mandelcorn, 1964, Academic Press New York and London.
O'Bryan H.M., Gallagher P.K., Brandle C.D., 1985, J. Am. Ceram. Soc., 68, 493-96.
Ohnishi N. and Iizuka T., 1975, J. Appl. Phys., 46, 1063.
Patent DDR DD 297 937 (C 30 B 15/20).
Peterson G.E., Ballman A.A., Lenzo P.V., 1964, Appl. Phys. Lett., 5, 62-63.
Peterson G.E. and Carnevale A., 1972, J. Chem. Phys., 56, 4848.
Peterson G.E., Carruthers J.R., Carnevale A., 1970, J. Chem. Phys., 53, 2436.
Prokhorov A.M., Kuz'minov Yu.S., 1990, Physics and Chemistry of Crystalline Lithium Niobate, Adam Hilger, Bristol and New York.
Rauber A., 1978, Curr. Top. Mater. Sci. 1, 481 (North Holland, Amsterdam, 1978).
Reik H.G. and Heese D., 1967, J.Phys. Chem. Solids, 28, 581.
Reisman A. and Holtzberg F., 1958, J. Am. Chem. Soc., 80, 35; 80, 6503-7.
Revkevich G.P., Ivleva L.I., Katsnel'son A.A., Kuz'minov Yu.S., 1979, Kristallografiya, 24, 1079.
Rosenblum B., Braunlich P., Carrico J.P., 1974, Appl. Phys. Letters, 25, 17.
Rossolenko S.N., Pet'kov I.S., Kurlov V.N., Red'kin B.S., 1992, Zh. Teor. Fiz., 62, 61-70.
Runyan W.R., 1965, in: Silicon Semiconductor Technology, McGraw-Hill, New York.
Schirmer O.F., Thiemann O., Wohlecke M., 1991, J. Phys. Chem. Solid., 52, 185-200.
Semiletov S.A., Bocharova N.G., Rakova E.V., 1987, Kristallografiya, 32, 189.
Shigematsu K., Anzai Y., Marita S., Yamada M., 1987, Japan J. Appl. Phys., 26, 1988-96.
Smyth D.M., 1983, Ferroelectrics 50, 93-102.
Smyth L.M., 1984, Prog. Solid. State Chem., 15, 145.
Slobodnik A., 1981, Surface Acoustic Waves, edited by A. Oliner, Russian translation, Mir, Moscow.
Svaesand L.O., Eriksrud M., Nakken G., Grandl A.P., 1974, J. Cryst. Growth., 72, 230.
Takagi K. and Fukazawa T., 1976., J. Cryst. Growth 32, 89.
Taylor D., 1990, Properties of Lithium Niobate, EMIS Datareviews Series No.5, 44.
Tokizaki E., Terashima K., Kimura S., 1992, J. Cryst. Growth., 123, 121-125.
Trauth J. and Grabmaier B.C., 1991, J. Cryst. Growth 112, 451-57.
Turner E.H., Nash F.R., Bridenbaugh P.M., 1970, J. Appl. Phys., 41, 5278-81.
Voron'ko Yu.K., Kudryavtsev A.B., Osiko V.V., Sobol' A.A., Sorokin E.V., 1987, in: Short Reports on Physics, FIAN, 2, 34.
Yamada Ts., Takemura H., Inoue Y., Omui T., Matsumara S., 1987, Japan J. Appl. Phys.,

26, 219.
Yariv A., Introduction to Optical Electronics, 1976, Holt, Reinhart and Winston, New York.
Zharikov E.V., Ivleva L.I., Kuz'minov Yu.S., Osiko V.V., 1975, Izv. AN SSSR, Neorg. Mater., 11, 875-879.
Zhong Gi-Gou, Jian Jin, 1980, in: Proc. 11th Intern. Quantum Electronics Conf. June 1980, IEEE Cat. N 80, Ch. 1562-0, 631.
Zhou Y.F., Wang J.C., Wang P.L., Tang L.A., Zhu Q.B., Wu Y.A., Tan H.R., 1991, J. Cryst. Growth, 114, 87-91.

Index

O

P

R

S

T

U

V